FISH CARS AND FISH CULTURE

RAILROADS AND SPORT FISHING IN AMERICA

R. W. HAFER

UNIVERSITY OF NEBRASKA PRESS · LINCOLN

© 2026 by the Board of Regents of the University of Nebraska

All rights reserved

The University of Nebraska Press is part of a land-grant institution with campuses and programs on the past, present, and future homelands of the Pawnee, Ponca, Otoe-Missouria, Omaha, Dakota, Lakota, Kaw, Cheyenne, and Arapaho Peoples, as well as those of the relocated Ho-Chunk, Sac and Fox, and Iowa Peoples.

For customers in the EU with safety/GPSR concerns, contact:
gpsr@mare-nostrum.co.uk
Mare Nostrum Group BV
Mauritskade 21D
1091 GC Amsterdam
The Netherlands

Names: Hafer, R. W. (Rik W.) author
Title: Fish cars and fish culture: railroads and sport fishing in America / R.W. Hafer.
Description: Lincoln: University of Nebraska Press, 2026. | Includes bibliographical references and index.
Identifiers: LCCN 2025022983
ISBN 9781496235145 paperback
ISBN 9781496246110 epub
ISBN 9781496246127 pdf
Subjects: LCSH: Fishing—United States—Economic aspects | Fishing—Transportation—United States | Railroads—Freight—United States | Women in fisheries—United States | BISAC: HISTORY / United States / State & Local / West (AK, CA, CO, HI, ID, MT, NV, UT, WY) | SPORTS & RECREATION / Fishing
Classification: LCC SH463 .H27 2026
LC record available at https://lccn.loc.gov/2025022983

Designed and set in Arno Pro by K. Andresen.

TO BOB ROWLANDS,
WHO ENJOYED FLY-FISHING
AND REALLY LIKED TRAINS.

CONTENTS

List of Illustrations — ix

List of Tables — xi

Introduction — xiii

1. A Brief History of Railroads in the United States — 1
2. A Brief History of Fish Culture — 23
3. The U.S. Fish Commission — 47
4. The Fish Commission, the Great Experiment, and the Railroads — 69
5. The Fish Car — 103
6. Railroads, the Landscape, and Sport Fishing in America — 125
7. Railroads, Women, and Fishing — 151

One Final Note — 171

Acknowledgments — 175

Notes — 177

Bibliography — 199

Index — 211

ILLUSTRATIONS

Following page 150

1. Comparative timetable
2. Meeting of the trains, Promontory Summit, Utah
3. Seth Green
4. Spencer Fullerton Baird
5. Robert Barnwell Roosevelt
6. Livingston Stone
7. Baird Station on McCloud River
8. Depositing fish in stream from milk cans
9. U.S. Fish Commission building with fish car
10. Interior of fish car
11. Grand Rapids and Indiana Railroad map featuring American grayling
12. Chicago Milwaukee & St. Paul Railway poster
13. Lackawanna Railroad ad featuring Phoebe Snow
14. Rock Island Railroad magazine ad for Colorado
15. Pullman poster

TABLES

1. Fish introduced into California, 1871–98 — 93
2. Data on use of fish cars, messengers, and scope of operations — 117
3. State fish cars — 120

INTRODUCTION

The idea for this book began more than forty years ago, while attending a Trout Unlimited meeting in my town of Kirkwood, Missouri. The speaker was Spencer Turner, a fisheries biologist with the Missouri Department of Conservation, and his topic was how rainbow got to Missouri: trout are not native to the state. He spoke about his historical research, including his tabulation of stocking records by the workers at the federal fish hatchery in Neosho, Missouri. These records date from the late 1800s. I thought it interesting enough that I asked for a copy of the report. He was nice enough to send me a hard copy; a PDF via email would be years in the future. I read it and shelved it. But I kept it.

Turner's report resurfaced during one of my office relocations. I read it again, this time thinking that there must be more to the story than stocking locations and numbers. My curiosity spawned several years of digging deeper into the story about how trout (and other fish) got to Missouri. That led to one book. But the more I looked the broader the story became. This led me to address the broader topic that is the theme of this book—namely, how the railroad industry combined with advances in fish culture, best thought of as the propagation of fish in artificial conditions, to forever change the nation's fishery and influence the development of what would become sport fishing in America. These intertwined stories will unfold over the course of the book. Before that, let me provide some perspective for why this combination of seemingly unrelated fields even came about.

The United States was rapidly expanding during the 1800s. The surging population portrayed a vibrant and growing economy. The factory-like cattle farms and ranches that we have today did not exist, so people relied heavily on nature as a source of animal protein. Fish, historically, was a primary source of such protein for many households. With what seemed to be an endless supply of fresh and saltwater fish, the stock was harvested with disregard for the longer-term consequences. So pervasive and persistent was the overharvesting that by the mid-1800s the once massive shad and salmon runs that East Coast inhabitants relied upon to meet their dietary needs began shrinking markedly. And even though overfishing would remain a problem for years to come, governments failed to adequately regulate commercial fishing. With no clear property rights over the fish of the sea, states were unable and, in some cases, unwilling to curtail individuals' right to catch as many fish as they wished. Commercial fishermen used devastating approaches in order to meet market demands. Given the depletion of the fish stock, something had to be done. This is where my story begins.

After several failed attempts by states to manage fish harvests, the U.S. government got involved. On a cold winter day in 1871, following days of debate, Congress voted to create the U.S. Commission of Fish and Fisheries (hereafter, the Commission). Soon after the Commission became a reality, the assistant secretary of the Smithsonian Institution and naturalist Spencer Fullerton Baird was appointed the first fish commissioner of the United States. Baird's Commission was given the daunting task of trying to figure out how best to stem or, even better, reverse the decline in the migratory fish stocks along the East Coast.

How important was creating the Commission? The *New York Times* made it front-page news. The entire story, which ran on 24 January 1871, was this: "Mr. Dawes, of Massachusetts, introduced a joint resolution for the appointment of a Commissioner of Fish and Fisheries of the coasts and lakes. Passed by 127 to 48." That was it. Little did the *Times* know that soon this start-up agency would undertake one of the most important and outrageous experiments yet tried in wildlife management. You see, under Baird's leadership the Commission would combine the latest advances in fish culture and railroad technology to try and manage the nation's fishery. It would begin with confronting the decline of migratory fish along the Atlantic Coast by transporting salmon from the Pacific Coast and transplanting them into eastern rivers. While they were at it, large numbers of these Pacific salmon also would be deposited into interior rivers to see if migratory runs to the Gulf of Mexico could be created. Over time the types of fish being relocated changed, but the one constant in all of this was the Commission's reliance on the railroad industry to get the job done.

You can probably tell so far that part of the story of how railroads influenced the development of sport fishing in America came about because they were helping the Commission with its plans to move fish across the country. The railroads' effect on the history of fishing turns out to be more than just hauling around the Commission's fish and its workers. Equally important, though largely overlooked in earlier work, is recognizing how important were the actions taken by the railroads themselves to increase the public's awareness of and involvement in all things outdoors. This, of course, included the novel idea of fishing as a recreational pastime and not merely as a means to put food on the table.

This part of the story takes two tracks. One track focuses on the actions taken by railroads that instilled in their passengers an appreciation of the outdoors. Railroad advertising often extolled the wonders of nature, wonders that it just so happens were most easily reached by taking the train. Railroads advertised the scenery along their tracks and, to make good on their claims, worked to protect the views passengers would see. A strong profit motive was at work, but that does not obviate their actions to protect and preserve the landscape along their routes and even the environment beyond their rights-of-way. If, as many believed, the outdoors had restorative powers for the stressed urbanite, railroads were the means to achieve this recuperation of the soul. Railroads offered the easiest way to reach the rustic retreat in the Poconos or, if rustic wasn't your thing, opulent hotels and lodges. Whatever your desired destination, railroads were the only sensible and practical way to get there, even if it was clear across the country.

The companion to this part of this story reflects on how railroads specifically encouraged women to engage in the emerging pastimes of tourism and recreational fishing. Throughout the second half of the 1800s, railroads adopted a number of innovations that made train travel safer and more comfortable. Even though all passengers benefited, much of the advertising of these changes focused on women. As the century was ending, society's behavioral standards were changing, and, with them, women in ever-increasing numbers were traveling by train.

With barriers to travel lowered, women took the opportunity to visit, among others, rural areas that the railroad companies promoted. If a company was marketing its service to, say, Colorado, it often featured a woman fishing, camping, or canoeing. Whether it was for Maine, Michigan, or Colorado, railroads often portrayed the era's "new woman" fly-fishing in some pristine stream. And even though it was all done to increase passenger traffic, railroads provided the emergent female outdoorswoman with the means and opportunity to engage in sports that were once the domain of men.

That's the gist of the story. How will I tell it? Here is a quick rundown of how I plan to proceed.

It would be silly for a book in which trains play such an important part not to start with an overview of the industry in the period examined. The opening chapter, "A Brief History of Railroads in the United States," is a truncated treatment of this massive topic. Instead of a detailed analysis I want to give you a sense of how and why the railroad industry began and then, over time, became essential to the economic and social development of the country. I do this by splitting the 1800s into two parts.

The first segment, which runs essentially from 1830 to 1860, deals with the railroad's emergent years. Aside from providing some basic facts (e.g., growth in miles of track, technological developments), an emphasis in this discussion is how essential the railroad became to society. Yes, it was seen as a boon to economic progress, moving people and goods farther and faster than ever before. Just as importantly, the development of the railroad industry had significant social implications. Proponents believed that the railroad would be a great equalizer: urban and rural, North and South would all become one connected nation as goods, people, and information flowed with increasing ease and rapidity. Others saw the railroad as a means to escape the increasingly undesirable features that defined urban living. Railroads played up their ability to transport the city dweller to the recuperative bosom of nature. This theme, which carried on throughout the rest of the century, aroused the public's interest in all types of outdoor activities, including fishing.

The second section of the chapter surveys the continued growth and influence of the industry from the Civil War up to 1900, my self-imposed end point for this story. The railroad system in the postwar era expanded rapidly, with more companies and thousands of miles of track being added every year. What dominates the history of this period is the completion of the transcontinental railroad—truly one of the most epic achievements to date. Now you could travel from coast to coast in a matter of days, not weeks or months. There was more to this half-century than that, however. Over time the industry became highly concentrated and rife with noncompetitive behavior. Mismanagement and corruption marred the public's opinion of the industry, leading to increased governmental regulation.

The opening of the transcontinental lines was a truly monumental feat that had far-reaching effects. An unintended consequence was that it enabled a group of individuals who had the radical idea that they could collect and ship Pacific salmon (and later trout) from Northern California to the East Coast, where they would be released in an attempt to replenish the declining Atlantic salmon spawning runs. Equally radical was the idea that eastern shad would go in the other direction to be transplanted into western waters. Through the web of railroads that crossed every state and now the country, the government mounted the most aggressive campaign to date of applying fish culture to manage the nation's fishery.

Wait, so what is fish culture? Answering that question is the purpose of chapter 2, "A Brief History of Fish Culture." Fish culture basically is the propagation of fish using artificial means: think fish hatcheries. Fish culturists are those who developed and honed techniques to maximize fish production, and to experiment with hybridizing fish to achieve desired characteristics (e.g., fast growth, hardiness). Fish culturists believed that their techniques could be used to increase the supply of fish, so much so that shortages due to destructive fishing pressure could be offset. The technology of fish culture may seem straightforward, but it was, like the railroad, an embryonic technology in the 1800s. This chapter places the development of fish culture in the United States into a historic context and explains why it plays a central role in this book.

The opening of chapter 3, "The U.S. Fish Commission," introduces you to the debate over the decline in coastal fish populations. The once teeming runs of shad and salmon along the Eastern Seaboard had plummeted, and people feared they would disappear altogether. Because fish were a dietary staple, explaining and rectifying the decline was a top issue for state governments and the public. Regulation seems like the obvious course of action, but, with neighboring states blaming each other for the problem and fishermen using different techniques blaming their counterparts, there was little success in finding common ground. The problem was so acute that by the end of the 1860s it was suggested that the federal government should investigate the issues and propose solutions. Following some debate, much of it heated, in 1871 Congress created a new agency called the Commission of Fish and Fisheries. Its charge was to study the coastal fish problem and make recommendations on how to solve it.

The story of how the Commission came to be highlights the economic and social significance of the so-called fish problem. The process by which

Spencer Baird came to lead the Commission is informative to understand the politics of the fish problem and the policies that the Commission would later undertake. You see, shortly after its creation the Commission was actively propagating salmon and shad in massive numbers and, with the railroads' help, transplanting them from coast to coast and to points in between. I refer to this as the Great Experiment.

In chapter 4, "The Fish Commission, the Great Experiment, and the Railroads," I survey the Commission's attempt to resolve the fish problem and the pivotal role played by the railroads. It is an understatement to say that the experiment was far reaching. The Commission built a hatchery in Northern California for the sole purpose of collecting Pacific salmon eggs, which, once fertilized, were transported by train clear across the country to hatcheries back East. Here they were hatched and the fry deposited in rivers and streams throughout the eastern half of the country. It was highly experimental, but success would be a boon to society. It also would cement the Commission's role as the agency that could effectively manage the nation's fish stock. But it amounted to more than just applied fish culture: moving fish eggs and fry about the country in unheard of numbers required the railroads to become indispensable partners.

Finding more efficient means of transporting ever larger numbers of fish from hatcheries to stocking locations leads to the focus of chapter 5, "The Fish Car." This chapter chronicles the development of the fish car—a railroad car designed specifically for the purpose of transporting fish—and how the Commission (and some states) used them to expand the limits of its ongoing experiment. The fish car represented the marriage of the latest in railroad technology and in fish culture. The fish car enabled the Commission to transport huge numbers of fish more effectively than ever before. Instead of moving thousands of fry-sized fish, many of which died soon after being planted, fish culturists could ship more and larger fish to deposit sites, thus increasing their odds of survival. The fish car era, which effectively began in the early 1880s, lasted well into the twentieth century until it was superseded by another new technology: the truck. To put its reign into some context, by the end of the 1800s the Commission's handful of fish cars were logging more than one hundred thousand rail miles a year. This extensive use of the fish car made the Commission even more dependent on the country's railroads.

The focus to this point has been on the evolution of the Commission's fish management policies and the help provided by the railroad industry. This activity underpins the development of sport fishing in the country: without the Commission's undertakings the angler in Iowa may never have had the chance to land a rainbow trout. (Rainbows are not native to Iowa.) But the railroad's role in this history amounts to more than hauling fish around the country. In the remaining chapters I consider other ways in which railroads influenced the development of sport fishing in America.

This line of inquiry begins in chapter 6, "Railroads, the Landscape, and Sport Fishing in America." Railroads were influential in developing the public's interest in the outdoors. Recall that in the earliest days of the industry railroads were thought to be the best means of getting urban residents into the invigorating and restorative outdoors. Railroads recognized that the landscape envisioned by passengers as their train rolled through the countryside was a selling point that they could exploit to boost ridership. Railroads considered the scenery an asset and worked to preserve it. In so doing they influenced their customers' appreciation for the outdoors: If railroads were concerned about the landscape, perhaps they should be, too? Because of their extensive network, railroads accommodated the public's desire to visit rural areas and engage in activities associated with the outdoors. Railroads advertised themselves as the most efficient way of getting passengers to the outdoors and the recreational activities it offers, like fishing.

A growing middle class during the later 1800s sought outlets for their increase in time available for leisure activities. One of these new outlets was tourism, or travel for travel's sake. The idea of going somewhere merely to visit or sightsee was a relatively new one, especially outside of the wealthier classes. Railroads thus became integral to the rise of tourism. To meet and enhance the demands of the tourist market railroads promoted the "discovery of place." Railroads could deliver interested parties to rural getaways like the camps in the Poconos or to family resorts in Michigan. Some companies even built popular vacation destinations of their own. What is important for my story is the fact that, in many instances, sport fishing became integral in the railroads' promotional campaigns. So, whether it was a posh hotel for the rich or a wilderness getaway for the adventuresome, railroads worked to satisfy the public's demand for vacation getaways. And, many times, this featured fishing.

Further evidence of railroads' support for sport fishing comes from the fact that many companies maintained streams and the fishing conditions along their tracks. A number of railroads even complemented the activities of the Commission by stocking the rivers and streams near their routes. Such actions not only helped sell tickets to individuals interested in fishing these locales but reinforced the Commission's goals of providing fish for food and as a recreational outlet for sport fishermen. What other industry could boast of such publicly beneficial endeavors?

I close the book with chapter 7, "Railroads, Women, and Fishing." Railroads employed numerous approaches to increase the number of women passengers. For example, some trains ran ladies-only cars. Many instituted specific rules for employees on how to treat female passengers, especially those traveling alone or with children. Railroads' efforts to boost female ridership coincided with the breakdown of what was considered "proper" Victorian behavior. Where once it would have been considered unthinkable, women traveling alone by train became acceptable.

Railroads encouraged this change in attitude with advertisements that often featured a woman riding alone in the comfort of a passenger car. Of course, she was enjoying the pastoral scene passing by outside the train window. Sometimes the woman in the ad was engaged in an outdoor activity, at some destination the railroad was promoting. Imagine a poster promoting, say, Yellowstone Park: the visual is a young woman catching trout in a stream with the caption "Come to Yellowstone for the Catch of a Lifetime!" Such marketing promotions not only cultivated interest among women to travel, but they also engaged their participation in outdoors activities, especially in fishing. The era of the outdoors woman was underway, and railroads were at its forefront.

In chapter 7 I also pose the idea that by railroads increasing women's interest in the outdoors they helped advance the emerging conservation and preservation movements. The discussion in chapter 6 suggests that railroads often took direct actions to preserve the wilderness areas through which their lines ran. Here I supplement that idea, but with a twist. As railroads were getting more women interested in travel and the outdoors, they increased the number of those concerned with the issues affecting the environment and wildlife. It has been suggested that if it wasn't for women's involvement, the movement to enact policies protecting the environment and wildlife would not have been as effective as they were. My suggestion is that this would not have happened if railroads had not afforded them the means and the motive to explore the outdoors.

My research has given me a totally new perspective on the evolution of sport fishing in America. The policies of the Commission (and its state counterparts) provided anglers with a much wider variety of species to catch across a larger geographical area than would have occurred naturally. Though some of these interventions have had debatable results, the effect on the evolution of sport fishing is undeniable. What surprised me—and now perhaps you—is the critical role that railroads played in this story. From the transportation and depositing of fish for the Commission to their own promotional campaigns that attracted more and more individuals, men and women, to this recreational pastime, railroads were indispensable to the development of sport fishing in America.

Fish Cars and Fish Culture

1

A BRIEF HISTORY OF RAILROADS IN THE UNITED STATES

It would seem remiss of me to write a book in which the railroad industry plays such a significant role and not provide at least some cursory discussion of its evolution during the period that will be my focus: the nineteenth century. I am sure that most readers have some knowledge about railroads in the United States, but my guess is that it focuses on the period following the Civil War, the time when train tracks crossed the plains and mountains to reach the West Coast. There is, in fact, much more. The early years of the railroad set the stage for how it became so integral to everyday life, to the expansion of the economy, and even to those who thought it a good idea to try and manage the nation's fish stock by moving fish around the country. Not to mention how in the latter decades of the century railroads were at the forefront of the emerging trend of pleasure travel—tourism. It turns out that railroads were instrumental not only in helping the government try to manage the nation's fish stock, but also in kindling the public's interest in both the outdoors in general and in the activities it offers, such as fishing.

The Early Years: 1830 to 1860

During the early 1800s the United States of America was growing and growing fast. With most of its population residing near the East Coast, it became critical to find the least costly and most efficient means of transporting goods—especially raw materials and agricultural products, from the then-western reaches of the country to the population centers in the East. Roads existed, but transporting merchandise and produce any significant distance along them using wagons and carts was not only expensive but also time consuming. To illustrate, around 1820 the cost of moving a ton of goods from Buffalo to New York City—about four hundred miles—took twenty days. Water routes along streams and rivers provided one alternative. So did man-made canals, perhaps the most famous of which is the Erie Canal.

The Erie Canal, completed in 1825, linked Lake Erie in the West to the Hudson River in the East. The canal allowed grain and other agricultural products to travel more quickly from the West to New York City, bolstering the city's prominence as the major East Coast port. Once opened the canal had an immediate effect on commerce: the cost of transporting that ton of goods from Buffalo to New York plummeted from about one hundred dollars a ton to only five dollars a ton. And now the trip was completed in only six days.

The success of the Erie Canal prompted the construction of canals in other states, including Pennsylvania, Maryland, Ohio, Indiana, Illinois, and Virginia. Those in state and local governments believed that funding such public works would boost their local economies. Even though great sums of money were invested in these projects, most were commercial failures.[1] Many of the canals built simply did not have the scope of business that would ensure their commercial success. For example, goods shipped along the Erie Canal to New York City served a much larger market—domestic and foreign—than, say, goods sent to Baltimore. Weather was another issue: many canals simply could not operate in the winter when cold weather froze over the canals.

Canals soon faced another major problem: the introduction of the newest advance in transportation, the railroad. According to one reviewer, "Their [canals] value had been almost entirely superseded by railways, which private enterprise soon constructed upon all their routes."[2] The impact of the first railroads was muted because the earliest railroads were constructed only for specific and local purposes. The Granite Railway of Massachusetts, for example, built in October 1826, hauled granite slabs to Bunker Hill for the purpose of erecting a monument. It covered a three-mile stretch between the town of Quincy and the Neponset River. Similarly, the Mauch Chunk railroad, constructed in 1827, moved coal from Pennsylvania mines to the Lehigh River for shipment. Though these represent the early beginnings of American railroading, a major caveat being that horses and mules were used to pull the cars along the track.

The Baltimore & Ohio Railroad was one of the first true railroads: it had a locomotive that pulled the cars behind it. To better compete with New York as an export center, Baltimore's civic leaders chartered the B&O in February 1828. The grand plan was for the B&O to run from Baltimore to Wheeling,

West Virginia. This would link Baltimore with the Ohio River and all of the produce that moved along it. Baltimore would thus be connected to the emerging agricultural production in the country's midsection. That dream, however, would take many years to fulfill. Several years after construction began, the B&O had managed to build only fifteen miles of track, all within the Baltimore city limits.

The other railroad company often considered to be one of the nation's first was the South Carolina Canal & Rail Road Company. Chartered in 1827, its purpose was to ship agricultural goods from inland farms and plantations that arrived at its western terminus in Hamburg, South Carolina (across the Savannah River from Augusta, Georgia) to Charleston. From there the goods would be exported to foreign markets. The South Carolina claimed several firsts in U.S. railroad history. It was the first to employ a steam locomotive built in the United States. Named the "Best Friend of Charleston," the locomotive was manufactured at the West Point Foundry in Cold Spring, New York. Another claim to fame is that on Christmas Day 1830, the railroad ran an excursion trip with 141 passengers, the most for a steam-powered train to date. Perhaps most importantly, when completed in October 1833 its 163 miles of track between Charleston and Hamburg made it the longest railroad in the world.[3]

The opening of other early railroads tells a similar story to the B&O and the South Carolina. The Mohawk & Hudson Railroad, for example, chartered in 1826 and completed in 1831, connected the Hudson River at Albany with the Mohawk River at Schenectady. Its route allowed shippers to avoid sending their goods down the Erie Canal. Another early railroad, the Camden & Amboy Rail Road Transportation Company, provided a link between the Delaware River near Philadelphia and the Raritan River which ran to New York City. When operations began in 1832, the Camden & Amboy's thirteen miles of track also offered a substitute to shipping goods via the existing canal system. It also represents an example of how Philadelphia tried to compete with New York for export business. And the Tuscumbia, Courtland & Decatur Railroad linked the towns of Decatur and Tuscumbia in Alabama. Both towns lie on the Tennessee River, and the railroad, established in 1830, offered shippers an overland route by which goods could be moved, a little more than one hundred miles.

Between 1830 and 1832, fifty-five railroad companies opened for business in seventeen states. A few western states, including Illinois, Indiana, and Michigan, claimed at least one chartered railroad each. The Northeast and New England states boasted the largest concentrations of population and commerce, which explains their lion's share of early railroad construction. More than half of the new railroads were in Massachusetts (four), New Jersey (five), New York (eleven), and Pennsylvania (fifteen).

It did not take long for railroads—and the cities that funded them—to push their operations into the rapidly growing Midwest. Recall that reaching the Ohio River and hence the agricultural producers of today's Midwest was a key objective of the B&O and those entities funding it. Railroads that connected Portland, Maine, with the British provinces created a conduit for goods to travel from the western states through the Great Lakes and down the St. Lawrence River. Similarly, as railroads spread west from Boston to Albany, New York, the expectation was that they would eventually link with the railroads of central New York and with those passing through New Hampshire and Vermont. These lines would give Boston, an archrival of New York City, a commercial link to Montreal, Canada.

All of this construction suggests, not incorrectly, that commerce was a key motivation to build a railroad. But railroads also offered a broader social effect that gripped the public's imagination. A common theme seized upon by early railroad advocates was the ability of railroads to bind the country together. The heretofore unmatched interconnectedness created by a rapidly growing network of track would enable individuals "to intermingle with greater facility," resulting in a "general advancement in knowledge."[4] With cities and towns and rural America increasingly connected by railroads, the likelihood that "the Union of the states would be indissoluble" increased.[5] Advocates even suggested that the railroads' "bands of iron," a popular euphemism for railroad tracks, created an "iron girdle" that would hold together the increasingly far flung parts of the country.[6] Advocates also believed that railroads would help democratize the country. Exposure through travel, it was suggested, would level the social differences between the urban and the rural, between the elite and the masses. The editor of the *American Railroad Journal and Mechanic's Magazine* even suggested that railroads "more than any other institution among us except our schools" would succeed in "placing people on terms of equality."[7] Railroads thus served as more than a means to advance commerce and economic growth: they also would advance societal cohesion.

Others supported expanding the railroad system because railroads offered the best means of escape from the urban areas. Even in those early times urban areas were disparaged as dens of ill health, corruption, and immorality. Railroads thus offered the downtrodden, harried urbanite the means of escape to rural America, idealized as an unspoiled wilderness with physical and mental recuperative powers. This notion that railroads would transport the urbanite to the healing powers of nature continued throughout the rest of the century. Indeed, a great deal of railroad advertising in the years following the Civil War promoted the countryside and the activities—such as fishing, hunting, and hiking—that it offered. As you will see, this function is important to understanding why railroads became integral to the growth of sport fishing.

The potential of the railroad captured much of the national psyche. There were, however, dissenters. Though largely ignored as reactionary twaddle, naysayers voiced concerns about the railroads' damage to the environment and its negative effects on societal relationships. Influencers like Daniel Webster dismissed these arguments, suggesting that the "serious" person correctly sets aside such misplaced concerns and accepts the railroad as the technological marvel that would propel the nation to greatness. Webster and like-minded supporters believed the railroad was such an important invention that it would truly alter the material and spiritual future of the country.[8] Railroads were considered to be a critical cog in the American industrial revolution that would, warts and all, ensure the nation's preordained position as the greatest economy in the world.

Modern localities use a variety of tax abatement methods to attract shopping malls or new businesses to build in their jurisdiction. Why? Mostly to lure jobs and tax revenues away from other cities and towns. Local politicians and city leaders in the 1800s played the same game with railroads: getting a train company to originate in your town (or just pass through it) increased the odds of economic success.[9] Many local and state governments thus backed privately operated railroads through the sale of government bonds or through direct loans. Some political entities even granted railroad companies exclusive rights: a clause in the Boston & Worcester Railroad charter prevented the construction of any competing route within five miles for thirty years.[10] Even with such protection from competition, a large number of railroads required constant infusions of capital to stay afloat.

Early enthusiastic support and funding of railroads was followed by a rather tardy conviction that the wrong thing had been done: it was not the business of a state or city to engage or assist in the construction of internal improvements.

Then came the full conviction that the states must release themselves from their industrial undertakings. In choosing between the two evils of going deeper into debt by continuing to financially assist railroads or of assuming, single-handed, as a debt, the aid already granted, states invariably chose the latter.[11]

Even with the numerous failures of railroads, new ones were popping up. There was also the constant merging of existing companies. Amazingly, regardless of poor management or lack of need for yet another railroad's service, the money flowed. Not surprisingly, this speculative investment in railroads led to significant losses. By the end of the 1830s, "an extraordinary and most unhealthy stimulus was given to the construction of railroads . . . by the wild and extravagant spirit of speculation which swept over the country."[12] The speculative bubble upon which railroad stocks were carried soon burst. Once a few railroads began to fail, investors in Europe—who bet heavily on the future of the railroad in the United States—began dumping their shares of U.S. railroads. The pervasive sell-off sparked a crash in the European stock markets that, as panic-driven financial crashes tend to do, spread quickly to the New York stock market. The ensuing financial panic resulted in the failure of many U.S. banks, a severe economic downturn, and a slowing in the pace of economic expansion that lasted well into the 1840s.[13] As it turns out, this was only the first episode of financial shenanigans associated with the railroad industry that would shake financial markets.

The financial panic and the depressed nature of economic activity in the United States wore on for several years. Even so, the nation's network of railroads continued to expand. By the end of the 1840s railroad mileage had more than doubled, from about 2,800 miles in 1840 to over 7,600 miles of track in 1849. Railroad transportation rapidly emerged as the most used method of shipping goods and moving people. The wave of expansion continued into the 1850s, with even greater increases in miles of track. With 7,600 miles of track in 1849, by the end of the 1850s it had more than trebled to nearly 28,000.

Much of this growth arose from the expansion in agricultural production in the Midwest and the rise of new urban centers located farther inland from the coast. For example, the New York & Erie Railroad sought to establish an unbroken trade route from the Mississippi River to New York City. Goods would travel north on rivers to ports in the Great Lakes, from there to Lake Erie, whence to be taken by rail to New York. The Pennsylvania Central Railroad, which ran from Philadelphia to Pittsburgh, hoped to eventually extend its reach to St. Louis and the Mississippi River. The proposed route would cross the states of Ohio, Indiana, and Illinois, thus allowing a number of smaller though growing cities in the Midwest to become trading partners with businesses in Philadelphia. I noted earlier that from its inception the B&O had planned to extend its track from Baltimore through Virginia to Wheeling, West Virginia, in order to access the Ohio River. It was not until 1852 that it accomplished this goal. From this western terminus the city of Baltimore gained access to goods and passengers throughout the Ohio River valley.

Even though the overall growth of the railroad system was impressive, it lacked coordination. One account advises that "if [one] assumes that the map [of the railroad system in the mid-1850s] shows the possible routes by rail for through movement of internal trade" when compared with more modern maps, "then his reading of the earlier map promotes error rather than understanding of the actual situation portrayed."[14] Maps showing railroad lines in 1850 look more like a bowl of spaghetti thrown at a map of the country than any attempt at some well-planned transportation system. This disjointed system was a product of unfettered capitalism: it almost seems that if you could raise enough capital, you too could start a railroad company. This helter-skelter railroad construction led to a significant number of mergers as smaller lines realized that their twenty- or thirty-mile road really wasn't economically viable. As some of the larger lines gobbled the smaller, less efficient ones, trunk lines formed on the more prosperous routes. Consolidation and the reduction of competition meant that some companies saw their market power grow. Indeed, as I discuss later, the abuse of this increased market power would be the bane of the industry in later years.

One principal factor supporting the continued expansion of the railway system took the form of federal land grants. The federal land grant was created by the Railroad Land Grant Act of 1851, sponsored by senators Stephen A. Douglas of Illinois and William R. King of Alabama.[15] The act granted six alternate sections (a section equals 640 acres, equivalent to one square mile) of land per mile of track to the Illinois Central Railroad and to the Mobile & Ohio Railroad, which connected Mississippi and Alabama. The Illinois Central Railroad received the first land grant. Land grants would become the funding lifeline for the transcontinental lines built after the Civil War.

The purpose of a land grant was twofold. First, they supplied funds for construction and covered the operational costs of the railroad to which they were made. As titular owners of the land, the railroads could sell the tracts to individuals along the proposed route. The provision of such government largesse often was the incentive needed to get a railroad built. The other purpose of a land grant was more social. By selling off the land, the railroads (and indirectly the government) encouraged westward expansion. One Illinois Central advertisement declared that by buying the land the purchaser would be located "near markets, schools, Railroads, churches, and all the blessings of civilization." Quite a claim, but it worked. One observer notes that "the Illinois Central was soon profitably selling thousands of acres of its land and bringing hundreds of new settlers into the state."[16] To those who opposed the grants, supporters countered that they were good for society: they not only encouraged railroads to build more track that moved more goods and people faster than before, but they also encouraged migration and could rapidly increase a state's population. What politician would deny his constituents the chance to enjoy the fruits of economic growth and development of their community?

Over time longer train routes, usually formed through mergers or acquisitions, would connect the East Coast with the western reaches of the country. Train tracks go both ways, so raw materials from the middle of the country flowed east to manufacturing hubs, and manufactured goods flowed from factories in the East to the increasingly populated western areas of the country. But it's important to recognize that it wasn't only raw materials and finished goods that flowed east and west: so did information. The expansion of the railroad system kept people across the still-young nation informed about the state of their union, not to mention trends in fashion and culture. Magazines and newspapers soon began to have a national readership, helping, as early advocates had envisioned, connect the increasingly dispersed population.[17]

Railroad expansion in the pre–Civil War era made this new transportation technology an emerging force in the country's economic and social development. Goods, people, and information were being shipped across this rapidly growing country in days, not weeks. Markets for those goods were expanding. And people were streaming westward, populating the central portion of the country with cities like Chicago, St. Louis, and Cincinnati growing in commercial importance. This stage of the railroad's development is impressive. But it was the linking of the two coasts that would firmly establish the confidence in the public's mind and the political rhetoric that America's manifest destiny was becoming a reality. The transcontinental railway opened a new era in economic development and in travel.

Railroads in the Post–Civil War Era: Reaching the Coast, Scandal, and Change

The colorful history of the railroad industry in the second half of the 1800s has been told many times, so my treatment is brief.[18] The second half of the 1800s witnessed a number of significant changes that improved the railroads' ability to rapidly move goods and people across the country. Indeed, it is why rainbow trout eventually came to inhabit nearly every state in the continental United States. The expansion to the Pacific Ocean also positioned the railroad industry squarely in the forefront of a trend in travel that was catching hold: tourism. As one railroad historian aptly summed up the period, "In the half-century after Appomattox they [railroads] would experience their golden age as the nation became the industrial giant we know today."[19]

It is interesting to note that the initial debate on the idea of extending a railroad to the far western shores actually began in the 1840s. But it would not be until the Pacific Railway Act of 1862 was passed that building the transcontinental railroad became a reality.[20] The act contained several key parts.[21] First, it created the Union Pacific Railroad Company, a privately run company funded largely by the federal government. Second, the act set out a proposed route that would begin from a point on the one-hundredth meridian of longitude west of Greenwich, England. Think of a north–south line that forms most of the border of Minnesota and runs south through eastern Nebraska, Kansas, and Oklahoma. The actual route that would run to the western border of the Nevada territory would be determined by the most practicable way. At its western point, the Union Pacific's tracks would meet up with the eastbound construction undertaken by the Central Pacific Railroad Company of California. Initially the eastern terminus for the Union Pacific was set about 250 miles to the west of Omaha, with branches reaching out to connect Omaha; Kansas City; Sioux City, Iowa; and Atchison, Kansas. These proposed routes satisfied parochial political interests but quickly proved problematic. After some debate Council Bluffs, Iowa, was chosen as the official eastern terminus of the line. In reality the Union Pacific headed west from Omaha.[22]

Lastly, the act authorized the aforementioned Central Pacific Railroad Company of California to build eastward from a site close to San Francisco to the California-Nevada border. Because the Union Pacific would be laying track over a much greater distance, it would be subject to more uncertain construction conditions and potential delays. The act thus allowed the Central Pacific to build farther eastward if it reached the California-Nevada border ahead of the Union Pacific. In effect the Central Pacific was given permission to continue laying track eastward until and wherever it met up with the westbound Union Pacific. The point was to get the thing built.

The incentive for each company was to lay track as rapidly as possible. For each mile of track laid, the government granted five alternate sections of land on each side of the track—that is, a swath of land ten miles wide on both sides of the route. This generosity was extended in the Act of 1864. That legislation increased the land grants to twenty miles on each side of the track. As with any land grants, the idea was for the two companies to sell off these parcels to fund construction and operation of their lines. As noted earlier, the government hoped that selling the land would produce the side benefit of encouraging settlement into the western expanses.

The two companies also received cash subsidies and long-term government loans to cover construction costs. The loans varied with respect to the terrain crossed: sixteen thousand dollars per mile for construction across the relatively flat plains to thirty-two thousand dollars per mile for construction in the more rugged plateau between the Rockies in the East and the Sierras in the West. The 1864 act also allowed the two companies to issue bonds in dollar amounts that equaled the government bonds.[23]

This grand plan to span the western half of the nation with a railroad stirred the public's imagination. Just think: someone in New York could travel to San Francisco in days, not weeks or months. The possibilities for migration and commerce were not lost on a Congress amenable to promoting economic growth. In the halls of Congress, "the raising of capital for the [Union Pacific] was to be splendidly simple and effective. . . . The national interest would be so keen [the capital stock] would sell itself, even though there were no surveys of assurance of earnings."[24] Such giddy optimism discounted the harsh realities of building such a large stretch of track through an untamed wilderness. Over time this unbounded optimism would be tarnished by the decidedly self-serving behavior of those managing the Union Pacific. More on that in a bit.

The Central Pacific, which incorporated in 1861, got the jump on the Union Pacific. Theodore Judah already had been lobbying state and federal officials for a railroad through the Sierra Mountains. He was unable to secure federal funding to support his idea, so he approached several influential and wealthy

Sacramento businessmen. These included Leland Stanford, a prominent businessman and the first Republican governor of California (1862–63); industrialist Collis P. Huntington; and merchants Mark Hopkins and Charles Crocker. Impressed by the commercial possibilities of owning the western portion of the transcontinental railroad—the westward pull of the recently discovered Comstock Lode loomed large—the so-called Big Four joined forces to incorporate the Central Pacific Railroad.[25]

Construction on the highly anticipated transcontinental railway began in early 1863, when the Central Pacific broke ground in Sacramento. The partnership between Judah and his backers soon soured, however, and construction slowed. Judah died of yellow fever that November, leaving the remaining partners to proceed as they saw fit. The Big Four "had not entered upon this thing through philanthropic or wholly patriotic motives ... [but] were hard-headed businessmen, inured to making their own way."[26] Their first change was to put one of their own, Charles Crocker, in charge of construction.

The lack of labor slowed the pace of construction. As a solution Crocker hit upon the idea of importing Chinese laborers to fill out the construction crews. Not only did they represent a large pool of labor, but they would work for less than domestic workers. By employing hundreds of Chinese laborers from Canton the Central Pacific made significant progress eastward, snaking its way through and across the Sierras. When it reached the initial objective of the Nevada line, Crocker approached Congress for authorization to continue building eastward through Nevada. Probably because he had demonstrated success thus far, Crocker got the authorization and the needed funding. The Central Pacific's crews were laying track at a pace unmatched by the Union Pacific. By 1869 the Central Pacific had reached the Utah border.

The story of the Union Pacific's construction is much less glowing. Back in Omaha ground hadn't even been broken by the end of 1863. In fact, it was not until the fall of 1864 that plans for construction across Nebraska and into Wyoming were finalized. Funding issues—public purchases of rail bonds proved difficult during the war years—and the scarcity of labor pushed construction by the Union Pacific back until July 1865. The fact that the Union Pacific got a late start was compounded by corruption and financial improprieties by the Union Pacific's funding company, Credit Mobilier. The financial improprieties were so rife that Peter A. Dey, the Union Pacific's first chief engineer, resigned in protest in 1866.

In Dey's place the railroad hired retired general Grenville M. Dodge to supervise construction. Dodge had an exceptional service record with the Union Army during the Civil War, and he brought his military style to the management of the Union Pacific's construction. Fairly quickly after Dodge took the reins the Union Pacific's track had reached North Platte, nearly three hundred miles west of Omaha. This progress was achieved with military-like control, and with army-sized manpower. At one point the Union Pacific employed about ten thousand construction workers together with equally impressive numbers of draft animals.[27]

As this army of men and animals laid track across Nebraska and into Wyoming, it began to meet resistance from Native Americans who viewed the railroad as an unwanted intrusion upon their land.[28] Even though the U.S. Army provided some protection, recurring raids interrupted progress. Even so, Dodge and his crews relentlessly pushed westward, extending the Union Pacific's tracks to 450 miles west of Omaha by the end of 1868. The overriding objective was to press westward, even if the cost was the quality of roadbeds and bridges.

The advance grading crews of the Central Pacific and their Union Pacific counterparts met up in Utah in early 1869. The problem was where to join the two lines. This question had not been addressed, and it seemed silly for each to continue building if it meant that the two lines would never converge. Following negotiations between the Central Pacific's Huntington, the Union Pacific's Dodge, and members of Congress, a deal was hammered out. In April 1869 it was announced that the tracks of the Central Pacific and Union Pacific would unite at Promontory Summit, Utah Territory.

And so it was that after years of construction, delays, and cost overruns, on Monday morning, 10 May 1869, two trains from the West arrived carrying Central Pacific officials and political dignitaries. From the East arrived a train carrying an entourage of Union Pacific executives, directors, and politicians. With bands playing, crowds cheering, and following perfunctory speeches lauding their accomplishment, Leland Stanford of the Central Pacific and Thomas C. Durant of the Union Pacific supervised the driving home of the now famous Golden Spike. With this celebration the transcontinental railroad was a reality.

Word that the two trains had met was telegraphed across a nation eagerly awaiting news of the monumental achievement. This news and the now-immortal picture of the two trains' cowcatchers touching galvanized the entire country. The completion of the transcontinental railroad was an accomplishment that everyone (maybe more so if you lived in the North than the South) could take pride in. And the new railroad was an instant hit. Only six days after the ceremonies at Promontory Summit, a train carrying five hundred passengers left Sacramento bound for Omaha. Soon there were two trains a day, in each direction, to meet customer demand.

Even though it was not inexpensive, passengers flocked to experience this new travel adventure. By the summer of 1869, the Central Pacific was moving nearly 30,000 passengers along its line. The Union Pacific claimed to have carried nearly 143,000 passengers during the next year.[29] Think of what this meant: taking the transcontinental railroad meant that a trip from San Francisco to New York only took about eight days. This was hardly conceivable, since only a few years earlier the same trip was much more perilous and took weeks if not months, depending on your mode of travel. What had been a fanciful idea was now reality: maybe those bands of iron really could link the country together.

The benefits of this extended rail system reached every corner of the economy. Even the cost of moving the mail declined.[30] Combined with the parallel expansion of the telegraph, news and information about all sorts of topics—cultural, political, and economic—presented in a variety of forms (e.g., newspapers, magazines) found their way across the country faster than ever before. It also made once un-heard-of ideas like transplanting fish from one side of the country to the other conceivable.

Along with the excitement of what the new transcontinental line offered arose the dark shadow of scandal. Corporate corruption was common during this period in U.S. economic history, but the extent of the scandal surrounding the Union Pacific's Credit Mobilier shocked even the most jaded observer. Credit Mobilier, as I noted earlier, was the financial management arm of the Union Pacific Corporation. It had a reputation of operating at the edges of the law, but the revelations that broke in 1872 were shocking. The *New York Times* described Credit Mobilier as "an inner ring of stockholders of the Union Pacific, who not only broke their contracts with the Government, but also looted the pockets of their fellow stockholders."[31] The Credit Mobilier scandal was, however, different from corruption in this era of unbridled capitalism. The public and political perception of railroads would be radically altered once the story broke wide open.

As financial manager of the Union Pacific, Credit Mobilier defrauded the U.S. government to an astounding amount. Credit Mobilier submitted invoices to the Union Pacific for costs of building, operations, and such at highly overstated values. The Union Pacific then submitted these inflated invoices to the government for reimbursement as per their agreement. The government and its watchdogs compensated the Union Pacific for what they believed (or chose to believe) were actual costs. Credit Mobilier then skimmed off the difference and distributed the ill-gotten gains to its directors and to others.

The depths of the corruption was publicly revealed (though one must believe that it must not have been a well-kept secret) in a story that appeared on 4 September 1872 in *The Sun*, a New York newspaper.[32] The story reported who was in on the scam and on just how much it had cost the government: nearly 44 million dollars. But the story was more than one of simple financial fraud. Remember when I said that the skimmed funds went to "others"? It turns out that among these others were a number of influential members of Congress, bribed to ensure favorable laws and rulings regarding the railroad's activities. Some congressmen were paid directly in cash; others were given the opportunity to purchase heavily discounted shares of Credit Mobilier stock which, when sold at market value, resulted in significant profits.

The magnitude of the fraud and the extensive bribery of public officials stretched the credulity of even those inured to corporate-political corruption at the time.[33] The newspaper story named governmental officials on the take. With public anger aroused, Congress had no alternative but to call for hearings. Out of this investigation the names of two serving congressmen and three sitting senators were submitted for discipline. Also caught up in the investigation were the sitting secretary of the treasury and President Grant's then–vice president and running mate in the 1872 election. In the end, though some reputations were sullied, few significant punishments were meted out. One prominent politician caught up in the scandal was Ohio Congressman James A. Garfield. He simply denied the charges that he received shares from Credit Mobilier and suffered no further damage. Why sort him out? For one, he will become the twentieth president of the United States. More importantly for my story, he will play a key role in the creation of the U.S. Fish Commission.

You might think that the Credit Mobilier scandal would have put major railroad projects into jeopardy. But, no, the allure and the importance of the railroad was so ingrained in the economy and society that railroad construction continued apace. In fact the Central Pacific–Union Pacific transcontinental railway was just the first of several routes to the West Coast. As part of the 1864 legislation, the Northern Pacific Railroad Company was given a massive land grant (roughly forty million acres) to build a railroad from somewhere on the western shore of Lake Superior west to Seattle. Under the financial management of the infamous Jay Cooke, the line rapidly pushed westward after its first rails were laid in February 1870 in Minnesota. After reaching Bismarck, North Dakota, however, a financial panic swept the country in 1873. The strained resources caused by the Northern's rapid construction pushed Jay Cooke & Company into bankruptcy that September. The Northern was in limbo until 1878, when construction resumed under new management and funding. After several years the Northern had laid track to west of Helena, Montana. Plagued by continuous financial mismanagement, it would take another decade before the Great Northern Railway, as it was then called, reached Seattle. From this sketchy beginning, this line will be a key player in the tourism movement in the last of the 1800s, being the only major railroad to serve what will become Glacier National Park.

From the early debates over the best route for the transcontinental railway there was agitation, mostly by Southern interests, to construct a route that supported Southern interests.[34] The Big Four of the Central Pacific wanted to build a branch line from Sacramento to Los Angeles and from there eastward. To achieve this they formed the Southern Pacific Railroad as a landholding company in 1865. By 1877 the Southern Pacific traveled south from Sacramento through Los Angeles and from there across Arizona, eventually connecting with the Texas and Pacific Railroad on the outskirts of El Paso. Because the Texas and Pacific Railroad originated in Shreveport, Louisiana, once a branch from New Orleans was constructed in 1883 this southern line linked the ports of New Orleans and San Francisco. Freight and passenger travel from the Deep South to California was open for business. Though more than a decade later than the original transcontinental route, the South finally had access to the West Coast, and to the tourist destinations that will develop in the Southwest.

By the end of the 1870s, the public's once rosy view of railroads had been severely tarnished, and not only by the Credit Mobilier scandal. Industry consolidation gave rise to large railroad companies that controlled freight and passenger traffic across significant regions of the country. Many companies used their market power to extract economic rent from their customers, often in the form of rate discrimination. This especially harmed western farmers, who paid higher rates relative to their eastern counterparts to move their products to the East Coast markets. According to railroad historian John Stover, "No longer a self-sufficient producer, but, rather, a man producing for a distant market, [the farmer] had no choice but to use the rail facilities offered at the rates ordained by a largely absentee ownership."[35] Some of the larger lines would charge the lowest rates for some long-haul shipments, offsetting any losses with higher rates charged to shippers or localities that did not enjoy such competitive pricing. They also engaged in rate wars that not only jeopardized the survival of smaller railroads but also made it difficult for shippers to calculate their costs. In the face of such difficulties, "consumers, farmers, and shippers endured rate discrimination and financial corruption.... They seemed... to agree [that] it was useless for men to stand in the way of steam engines."[36]

I mention this abuse of market power because later you will find that the U.S. Fish Commission and its state counterparts enjoyed special treatment from most of the railroads with which they did business. The commissions publicized their activities in annual reports and included glowing accounts of how this or that railroad company provided transportation for personnel or materiel at below-market rates, sometimes for free. This in-kind support was even extended to those commissions who in later years employed their own specialized train cars to ship fish. Why did railroad companies offer such special treatment to the commissions? I'll leave any speculation on that question for a later chapter.

The industry's widespread abuse of market power did not go unchallenged. Grange interests effectively lobbied state legislatures in Illinois, Iowa, Wisconsin, and Minnesota to pass laws that barred the railroads' pricing schemes. Some states established railroad commissions to set maximum railroad freight rates in their states. In the face of mounting anti-railroad sentiment, most state governments adopted a decidedly laissez-faire approach to dealing with the railroads.[37] A not uncommon view was that "no government is capable of properly executing or administering *commercial* enterprises, and our own [government] has wisely committed all such matters to the *people*, where they belong" (emphasis in original).[38] It was best for business and government to remain in their own lanes of influence. Or so it seemed.

Railroads responded to this affront to their power by successfully pressuring many of the state legislatures to reverse laws having to do with their ability to set freight rates. Even so, the public's attitude toward the railroads had soured and was not to be dissuaded. Anti-railroad sentiment in Congress was increasing, illustrated by the thirty or so separate bills regulating railroads introduced between 1874 and 1885. In 1877 the U.S. Supreme Court reaffirmed the states' right to regulate railroads and the Act to Regulate Commerce, better known as the Interstate Commerce Act (ICA) was passed. The ICA made illegal the practice of discriminatory pricing between high- and low-volume freighters through the use of rebates and the charging of lower rates for long hauls compared to short hauls. It also created a federal regulatory commission to oversee its application: the Interstate Commerce Commission (ICC).

These regulatory changes seemingly had little effect on the railroads. What did impact the industry was the financial crisis of the late 1880s. The crisis, along with overbuilding and cutthroat competition, stressed the financial foundations of many railroad companies, pushing dozens into bankruptcy. The larger survivors turned to investment banks and reorganized their finances. Out of this activity arose seven "communities of interest," as J. P. Morgan called them. These seven companies controlled shipping on nearly two-thirds of the country's track. Analogous to the behavior of more recent oligopolies like the OPEC oil cartel, the oligopoly of railroad companies wielded enough market power to control the flow of goods and set rates. Like other industrial behemoths of the time, the railroads would be caught up in the federal government's trust-busting actions of the Progressive Era (1900–1917). The Elkins Anti-Rebate Law of 1903 allowed the government to intercede in interstate shipping rates. More significantly, the Hepburn Act of 1906 expanded the jurisdiction of the ICC, allowing it to set maximum railroad rates. It also gave the ICC access to railroads' financial records. It even extended the ICC's power to regulate covered bridges, terminals, and sleeping cars. The days of the freewheeling railroad company were over.

Innovations

Before leaving this brief history of the railroads, it would be useful to recognize some of the innovations that occurred. Although there were too many to fully account for here, I'll identify a few that helped influence the railroads' part in the development of sport fishing.

STANDARDIZING TRACK

A major hurdle to moving goods and people efficiently was the fact that different railroad companies often used track of different gauge (width). For example, when trains of two companies using different track gauge met in some town, transfer devices were needed to move freight between the lines' cars. Sometimes this entailed actually lifting cars from the chassis of a train fitting one track gauge and moving it to the chassis of a train that fit another. One observer lamented that "the many transshipments of freight are known to be the chief cause of delays, overcharges, and damages, besides adding greatly to the working expenses in labor and clerk-hire and requiring large and costly accommodations for the performance of the service."[39] Why did different track gauges persist? The best answer is inertia: companies that began with one track gauge were unwilling to convert to another. After all, refitting all of your track to a different gauge was quite costly and time consuming. Having different gauge track also was a way to maintain market power over rail traffic in their region.

The move to standardize track gauge got a boost with the Pacific Railroad Act of 1862. The act left it for President Lincoln to decide what the gauge of the transcontinental line would be. After consultation with experts, he chose 5 feet. Lincoln's decision sparked a heated debate that laid bare the commercial interests and political might of Northern railroad companies, the vast majority of which happened to use a gauge of 4 feet 8 ½ inches. To end the debate, Senator James Harlan of Iowa sponsored a bill in 1863 making the gauge of the Pacific Railroad 4 feet 8 ½ inches. Harlan's bill passed overwhelmingly in both the Northern dominated Senate and House.[40] This "standard" gauge was widely adopted, and by 1890 "only a negligible percentage of the country's railroads were of other than standard gauge. Rolling stock, both passenger and freight, increasingly equipped with standardized coupling and braking equipment, moved smoothly from line to line. Physical obstacles to the free flow of traffic had been largely eliminated."[41]

TECHNOLOGICAL INNOVATIONS

Let's start at the front of the train. In the 1830s the standard locomotive had six wheels and weighed about ten tons. It was sufficient to pull a few cars over fairly flat terrain but proved inadequate as the railroad system expanded and its uses became varied. So, locomotives got bigger: the average locomotive by 1850 beefed up to twenty tons and rolled on eight wheels. Ten years later the typical engine weighed in at twenty-five tons. The largest locomotives were forty-two tons and, more importantly, were able to take curves safely at speed, a critical improvement if trains were to include more cars. Originally used by lines hauling raw materials, the larger locomotives became more common, hauling heavy freight or long trains over great distances of varied terrain.

The capacity of freight cars increased as railroads became the dominant method of moving merchandise long distances. The average freight car in 1850 measured between twenty-four and twenty-eight feet and carried eight to ten tons of cargo. The common freight car in 1880 was thirty-four feet long with a twenty-ton capacity. Soon freight cars were built to haul twenty-five to thirty tons of freight. Using larger freight cars significantly lowered the costs of shipping goods across the country.

A number of innovations focused on safety improvements. Because trains often traveled in open areas, free-roaming cattle and wildlife presented obvious hazards. The solution was to add pilots (cowcatchers) to the front of engines. And—though you may think this odd—bells and whistles and headlights to illuminate the track and warn oncoming traffic which were not there in the early days of railroading became commonplace over time.

The passenger car evolved as railroads competed for passengers.[42] A typical passenger car in 1840 was thirty feet long and less than nine feet wide. With seats on each side, this meant that the center aisle was only eighteen inches wide. With a capacity of about fifty passengers, traveling in such a car was cozy, to say the least. (It was still preferable to alternatives, like the stagecoach.) The typical coach car by the 1850s had lengthened to forty feet, was a bit wider, and held more passengers who rode on wooden benches. As passenger travel increased, additional amenities began to appear. This included raising the central roof and adding windows in the roof to allow for more light and ventilation. The latter proved to be a two-edged gift: the windows let more air in, but this sometimes meant more soot and smoke infiltrating the cabin. Over time lavatories were installed in passenger cars, as well as stoves to provide heat in the cold months.

A significant trend of improving the passenger car—and the general travel experience—began in the 1860s. Perhaps the most notable is the sleeper car. The Pullman Palace Car Company of Chicago was not the only company to manufacture sleeper cars, but its cars soon made it the dominant company in this market. In addition to its popular sleeper car, it heightened the luxury of train travel by introducing its hotel car in 1867. The hotel car combined sleeping arrangements and meal service. A year later it began producing a line of diner cars, the most famous of which was the Delmonico, named after the famous Manhattan restaurant. Pullman dominated the business, and the public's imagination. By 1900 railroads were moving more than 3,200 Pullman cars across the country, annually transporting over 7.5 million passengers.

A number of safety improvements to make train travel safer also were adopted. The air-brake system patented by George Westinghouse was a significant improvement. Westinghouse brakes were much more efficient at stopping a train, requiring only one brakeman, rather than the one-per-car brakeman that was needed before. Other notable changes that enhanced the travel experience included interior lighting and heating, and the addition of vestibules at the ends of the cars. These and other changes improved the passenger experience and, in doing so, heightened the public's interest in using the railroad for all types of travel, business and pleasure. Such innovations were important in growing that part of the business related to tourism.

Railroads competed for passengers and in doing so often used advertisements that showed women traveling alone or with children. This, it turns out, was one way by which railroads influenced the nascent tourist industry and, by extension, the development of sport fishing. By promoting the idea that train travel was safe enough for women to travel alone—remember, these were Victorian times, and such behavior was not considered acceptable for a respectable young lady—the railroads promoted the idea that everyone should and could visit rural America and participate in recreational activities that had once been the purview of a few. Indeed, this was, a half-century later, the very same position taken by the early railroad advocates.

STANDARDIZING TIME

I've saved this for last because it truly demonstrates how embedded the railroad was in daily life. "The pulse of on-line communities [those with railways running through them] beat in 'perfect time' with that of the railroad," as one observer puts it.[43] This "pulse" was the mental timekeeping that arose from the predictable arrival and departure of the train. For travelers and shippers, however, the problem was that not all train companies used the same time. That is, what was one o'clock for one railroad might not be one o'clock for another.

Time discrepancies in railroad schedules existed because noon in a railroad's schedule was the time when the sun was at its highest at that company's home office. Train schedules were, therefore, complicated affairs. Travelers and shippers solved this problem using publications that included what one called its "Comparative Time-Table." Such a table appeared in Vernon's 1868 *Official Railway Guide of the United States and Canada*. The table listed the local time at ninety different cities using noon in Washington DC as the basis of comparison. Along with the departure and arrival times for various railroads across the cities and towns they served, passengers by "easy calculation [determined] the difference in time between several places" to coordinate their travel itinerary.[44] Even with such aides, making connections during a long-distance trip surely was no simple feat.

An industry group known as the Time-Table Convention and the General Time Convention met periodically throughout the 1870s to discuss solutions to the time problem. At the 1883 meeting of the group, William Frederick Allen, an employee of the Camden & Amboy Railroad and the secretary of the First General Time Convention, submitted a proposal to divide the country into four time zones. They would be delineated by, respectively, the 75th, 100th, 105th, and 120th meridians west of Greenwich, England. After some debate it was agreed that the railroad industry would adopt Allen's time zones. And so it was that at noon on 18 November 1883 the railroad industry adopted Allen's proposed time zones and created what was then called "railroad time." Once railroads adopted these time zones in making their schedules, the public fell into line. Interestingly, however, it wouldn't be until 1918 that Congress got around to passing the Standard Time Act, establishing railroad time as the official time of the country. Of course, Allen's four zones are today known as the Eastern, Central, Mountain, and Pacific Time zones.

A Final Note

From a handful of companies operating on a few dozen miles of track in the 1830s to an industry of over one thousand companies with nearly two hundred thousand miles of track by the end of the century, the railroad industry had become a colossus of the American economy. The railroad became integral to nearly every aspect of American life. From moving freight and people to influencing how we tell time, railroads made their presence known.

As railroads became indispensable to American life, they would prove equally important to those who believed they could assist on reversing the decline in the nation's stock of fish. With stocks of coastal fish like shad and salmon in notable decline, some believed that the application of fish culture—the use of artificial means, like hatcheries, to produce far more fish than relying on natural reproduction—could reverse the declining fishery in ways that nature could not. If fish culturists could raise fish in hatcheries and enlist railroads to transport them in huge numbers, might spawning runs be saved? Could the rivers and streams depleted of once teeming fish populations be revived? Those questions, and more, are the subject of the remainder of this book.

2
A BRIEF HISTORY OF FISH CULTURE

Different forms of fish culture have been practiced for centuries. People in ancient China, Rome, and elsewhere caught wild fish and kept them in ponds or manmade pools as a source of food or entertainment. What is different about fish culture as it developed during the eighteenth and nineteenth centuries is that it became a scientifically based approach by which humans could intervene in the natural reproductive process of fish. Fish culturists found that they could improve upon nature's success rate in the fish eggs-to-adult equation. As fish culture became more and more sophisticated, practitioners found ways to create hybrid varieties of fish with certain desired characteristics: faster growth, hardiness to water temperature changes, or even spawning times. Robert Barnwell Roosevelt, the one-time congressman, avid promoter of fish culture and, yes, uncle to the future president, suggested as far back as 1873 that "what was done with common tomatoes, potatoes, onions, and hundreds of other vegetable[s] ... may in a higher degree be carried into effect with fish."[1] Fish culture could make fish a food source just like chickens and cattle. More importantly, fish culture could hold the key to solving the ongoing decline in the nation's stock of fish.

Decline of Fish Stocks

Fish had long been a seemingly unlimited source of protein in many societies. It is hard to imagine today, but fish were at one time so plentiful that people complained about eating them. In his 1868 history of fish culture in America, W. F. G. Shanks observed that

> it is only a couple of centuries since it was the custom with Scotch and English house servants, in renewing their agreements with their masters, to stipulate that they were not to be compelled to eat salmon more than twice a week;

about the same period servants in Roman Catholic countries plead against the introduction of fish on the table on other than fastdays; and it is not a hundred years since the State of Connecticut, by a duly enacted law, prohibited masters from forcing trout on their apprentices oftener than three times a week.[2]

Because the supply of fish far outpaced the demand for it, fish were considered almost a costless commodity. And that is how they were treated. Noted British writer James Bertram writes that

> most of our public writers who venture to treat the subject of the fisheries proceed at once to argue that the supply of fish is unlimited, and that the sea is a gigantic fish-preserve into which man requires to dip his net to obtain at all times an enormous amount of wholesome and nutritious food. I would be glad to believe in these general statements regarding our food fisheries, were I not convinced ... that they are a mere coinage of the brain.[3]

With fish considered an unlimited resource, and because no one held property rights to them—except, of course, for the fish on royal property or those swimming in private streams—their once teeming numbers declined dramatically in England, many European countries, and America.

Within a relatively short time, the scarcity of fish, not their abundance, became the concern of many. It has been estimated that annual pre- and early colonial spawning runs of Atlantic salmon in eastern Canada and northeastern United States totaled 10 to 12 million fish, a biomass amounting to over 14 million pounds. The biomass of the annual spawning run of Pacific salmon along the West Coast is estimated to have been even bigger: around 352 million fish amounting to nearly 500 million pounds.[4] And yet by the early nineteenth century the take from the Atlantic salmon runs in New England already had noticeably fewer fish, and the distribution of different species in its southern range of the run had declined. The decline only deepened over time. In its assessment of the nation's fisheries in 1860, the U.S. Census reported that salmon "have now nearly forsaken the Merrimack, the Cumberland, the Thames, the Hudson, the Susquehanna, the Delaware, and other Atlantic rivers of the United States in which they were formerly found and taken in considerable numbers. Few are now caught south of the Kennebec."[5] The Pacific salmon faced similar declines once commercial fishing companies set up cannery operations along the West Coast. Salmon were by no means the only overfished species. The spawning runs of eastern shad in East Coast rivers at one time were huge. In a reminiscence written in 1881, one individual recalled that in his youth the shad run in the Susquehanna River near Scranton, Pennsylvania, was so large that "from the banks of the river at this fishery could be seen great schools of shad coming up the river when they were a quarter mile distant. They came in such immense numbers and so compact as to cause or produce a wave or rising of the water in the middle of the river extending from shore to shore."[6] Descriptions of the rivers during the shad spawning runs described them as "running silver." Even so, the 1860 census quaintly described the problem by observing that shad were "a timid fish" and thus "has become less plentiful than formerly."[7]

As the population of the United States increased—and it increased quite rapidly—so did the demand for fish as a part of the daily diet. According to the U.S. Census, the U.S. population more than doubled between 1800 and 1830, increasing from a little over 5 million to nearly 13 million, respectively. Over the next twenty years the population almost doubled again, reaching slightly more than 23 million by 1850. The growth in urbanization also was considerable. The percentage of the population designated as living in an urban area rose from about 6 percent in 1800 to over 15 percent by 1850. To pick just one example, the population of New York City rose eightfold between 1800 and 1850, from around 60,000 to more than 515,000. With most of the population living in urban areas, dietary demands pressed heavily on the mostly wild sources of protein, such as the nation's fishery.

The Fish Problem

The fish problem, as it was sometimes called, was so acute that some of the best minds of the day turned their attention to finding an explanation. George Perkins Marsh was one of the earliest to offer an assessment and possible solutions. Born into a politically active upper-class Vermont family, Marsh attended Dartmouth and studied law upon graduation. He was admitted to the Vermont bar at age twenty-four. Marsh engaged in local politics at the local and national levels, was elected to Congress, and served as one of Vermont's representatives from 1843 to 1849. He would serve as a foreign emissary in President Zachary Taylor's administration. In addition to his political activity, Marsh reportedly spoke over twenty languages and wrote a dictionary of the Icelandic language as well as a biography of the camel. On top of all that, he is best remembered for his early work in conservation. His book *Man and Nature*, first published in 1864, is still in print today and represents one of the first U.S. works to specifically consider the effect of modern society on nature.

The Vermont legislature commissioned Marsh to write a report on the present state of the artificial propagation of fishes. Some in the legislature were curious about this new idea and whether it could be used to stem the decline in salmon and shad populations.[8] This curiosity did not arise from some ecowarrior mentality. Fishing was a major industry in Vermont (and other New England states), so fiscal realities drove the desire to solve the decline in the fishery. Marsh produced his report in 1857, the tone of which presages much of his argument in *Man and Nature*.

Marsh recognized how the forces of progress affect the natural environment. His proposal for how to deal with this may seem odd to the modern reader. Marsh seemed to take the degradation of nature as a fait accompli, writing that "we must, with respect to our land animals, be content to accept nature in the shorn and crippled condition to which human progress has reduced her."[9] Marsh and others believed that the forces behind economic growth and progress were simply too powerful to counter: the political will to establish and enforce regulations to protect the environment and wildlife from America's economic and societal advance was a rare commodity.[10] In essence, states thought themselves powerless to regulate the taking of fish and game by their citizens or by commercial hunters and fishermen.

Most of the population was loath to abide by any hunting or fishing regulation: nature's bounty was put there for the taking, right? Marsh thus believed that it would be nearly impossible to stem the decline in wildlife through the use of regulation. Not so for fish, however. The next sentence to that quoted above provides a ray of hope: "We may still do something to recover at least a share of the abundance which, in a more primitive state, the watery kingdom afforded."[11]

How were fish to be spared from the onslaught of civilization's onward march? The answer lay in the fact that recent developments in the relatively new (at least in the United States) field of fish culture might just provide a "scientific" solution to the degradation of the fishery. Marsh suggested to the politicians in Vermont (and to anyone else interested in the problem) that if fish culture could be conducted at the appropriate scale, many species of fish could be artificially propagated and sold in the marketplace as a substitute for wild fish. Perhaps more importantly, these fish also could be raised in sufficient numbers to stock depleted streams. Marsh held out the hope that fish culture could restore the salmon and shad runs of yesteryear. Marsh concluded his report with the observation that the underlying problem affecting the fishery "is everywhere a condition of advanced civilization and the increase and spread of a rural and industrial population."[12]

Robert Barnwell Roosevelt was another early and vocal participant in the debate. Roosevelt, whom you've already met, offered a historical perspective to explain the reduced numbers of fish, especially trout, in some of New England's most revered streams. With improvements in transportation, Roosevelt decried, came "so-called sportsmen [who] poured over the country in myriads, following up every rivulet and ranging every swamp, killing without mercy thousands of trout and hundreds of birds, boasting of their baskets crowded to overflowing, and counting a day's sport by the hundred; till Bashe's Kill, where the pearly-sided fish once dwelt abundantly, was empty, and the broad Mongaup, the wild Callicoon, and even the joyous Beaver Kill, with its innumerable tributaries, were exhausted."[13]

Roosevelt and others blamed railroads for expanding into once remote areas and providing more egalitarian access to fishing meccas once the purview of mainly the rich. But the general lack of a proper sportsman's ethic represented an equally important cause of the decline of wild fish and game. Marsh pointed out that the "diminution of the fish is generally ascribed mainly to the improvidence of fishermen in taking them at the spawning season, or in greater numbers at other times than the natural increase can supply."[14] Roosevelt disdained those for whom the goal of fishing seemed to be catching (and killing) as many fish as one could. Like others in the nascent conservation movement who sought to protect fish (and wildlife in general), it has been suggested that Marsh, Roosevelt, and others in their socioeconomic strata suffered from severe class distinction. Loss of their once privileged access to the famous streams of the East caused alarm. Even so, the basis of their apprehension was spot-on. Easier access via the expanding railroad network to streams with finite populations of fish and few if any regulations made the outcome inevitable. "Pot-hunters" and "game hogs," as those who fished with little regard to the stock of fish were often called, exemplified "the greed of man [which] when thoroughly excited, can extirpate, for mere immediate gain, any animal, however prolific it may be."[15]

Such undesirable behavior was a general problem. Captain Daniel Bradley recorded in his diary that soldiers from a garrison along the Great Miami River in Ohio "caught 2,500 weight of fish and about as many [the next night].... We have more fish than the whole garrison can make use of."[16] George Jerome, the fisheries superintendent of Michigan (1873–79), lamented that "anglers from abroad, and home beaters of the rod visit the haunts of the Trout and Grayling [in Michigan], succeed in large takes, eat a few, bag a few—and the great balances are left to rot on the beach in the summer's sun."[17] Indeed, the image of a successful fishing trip (or hunting trip) all too often, and for too many years, is a group of grinning fishermen standing behind stringers with dozens of fish. Too often the measure of success was (and still is) quantity, not quality.

Other factors also were negatively affecting the fish stock. Commercial fishermen who used any means necessary to meet the insatiable maw of commercial markets were a popular target of the conservationists. And not without cause. Commercial fishermen often dynamited a stream or a lake to maximize the harvest. After the explosion they would cull out the desired fish and leave any other fish that boiled to the surface to rot on the bank. The unrelenting drive of industrialization also had major consequences. "The erection of sawmills, factories and other industrial establishments on all our considerable streams," notes Marsh, "has tended to destroy or drive away fish, partly by the obstruction which dams present to their migration, and partly by filling the water with [pollutants] which render it less suitable as a habitation for aquatic life."[18] To provide perspective, in Pennsylvania's Lancaster County alone, 147 milldams were constructed on tributaries to the Susquehanna River between 1756 and 1776. Over 330 additional similar obstructions were built between 1805 and 1817.[19] Such unchecked advances in industrialization devastated both the salmon and shad spawning runs in the Northeast.

Agricultural practices of the times were harmful to fish habitat as well. A modern assessment of the situation is that "the wholesale transformation of this landscape from prairies, forests, and wetlands to agriculture and urban areas along with logging of the old growth forests was a significant contributing factor to the decline in the continent's fisheries legacy."[20] Clear-cutting forests to increase arable land increased the variability of water flow. During times of high rainfall, without the forest to diffuse the flow, water and sediment more easily poured into streams. Major flooding often occurred where none had before. This variability of water flow changed the ecology of the streams: floods often resulted in loss of food sources for fish and other stream inhabitants, and they affected breeding if the floods occurred during spawning season. Forest removal even altered the temperature of streams, making them less inhabitable by some fish, especially cold-water species like trout.

All of these harmful effects were visible to anyone interested. But the fact is that achieving economic greatness—realizing the country's manifest destiny—became an overriding goal. Even the staunchest of conservationists were not unaffected by this common aim. Marsh would write that "we cannot destroy our dams, or provide artificial water-ways for the migration of fish, which shall fully supply the place of the natural channels; we cannot whole prevent the discharge of deleterious substances from our industrial establishments into our running streams."[21] Another example is the observation made by the influential outdoors writer Charles Hallock during a fishing trip to Northern Michigan. On his trip he encountered the dredging of a stream to better connect Petoskey's Little Traverse Bay with the town of Cheboygan. The dredging resulted in sand being thrown out on either side that "is kept from drifting into the stream again by long rows of piles and planking." He shrugged off the effects of the dredging on the stream even though this dramatically altered the nature of the stream and made the bottom of the stream irregular. He shrugged this off, content with the fact that after the dredging "from the deepest holes one can take bass with a fly while standing on the bank."[22] Another writer mourned that "it is sad to contemplate the extinction of the 'angler's pride' [brook trout] in public waters.... The stern fact remains that in this utilitarian age its [the trout's] days are numbered and its fate irrevocably sealed."[23] No call for change, just the resigned observation that if the price paid for progress was the extermination of the brook trout, so be it.

If the effects of industrialization and economic progress could not be checked, why not impose regulations to at least stem the effects of overfishing? At the time and for many years to come the efficacy of fishing and hunting regulations was severely limited. Marsh offers the popular assessment that "the habits of our people are so adverse to the restraints of game-laws ... that any *general* legislation of this character would probably be found an inadequate safeguard."[24] Some felt that regulations (e.g., number of fish, minimum size limits, fishing seasons) simply prevented the ordinary citizen from pursuing their God-given right to harvest fish (and other game) which nature provided. Besides, even if enacted—and there were some fishing regulations in some states—the difficulty of enforcing them (e.g., too few wardens, no budget for enforcement, unfriendly courts) often rendered the topic moot.

If fishing regulations could not effectively reduce overfishing, why not try to limit access? This private sector approach explains the increased number of fishing clubs. "Wealthy sportsmen, appalled by the destruction of good sporting grounds," writes Paul Schullery, "simply bought up their own 'preserves' and kept everyone else out."[25] The exclusionary approach to wildlife management was quite unpopular, needless to say. A story about opening day of the New York trout season in the widely read *Forest and Stream* maintained that the "practice of establishing private reserves cuts off the general public more and more inexorably from the desired trout waters." The article even suggested that such behavior "will be a nice question of law."[26]

Ironically, the very fishing clubs that this *Forest and Stream* article denounced actually were protecting the fish stock by setting up and enforcing property rights over fishing grounds. Beginning mostly in the East, fishing clubs popped up further west as the population (and wealth) moved in that direction. As storied as some of these clubs were, their limited reach made them ineffective in quelling the general damage from overfishing. Still, through their example and their often well-connected membership, fishing clubs and game preserves helped sow the seeds for the conservation movement that would gather steam as the century wore on.

How to solve the problem of overfishing? If not regulations or privatizing the waterways, then what? As much as the country's economic destiny took prominence in public discourse, so too did the Victorian mindset that nature could be shaped to satisfy human needs. If economic progress and population growth resulted in demands on the fishery that exceeded nature's ability to meet them, then the answer seemed obvious: apply modern science and the emerging technology of fish culture.

Fish Culture to the Rescue

The raising of fish, for food and for pleasure, has been practiced for centuries. But *fish culture* is much more involved than *fish husbandry*: the raising fish from fry to adulthood or tending to a stock of fish that had been transplanted from one location to another. The emerging practice of fish culture had a practical and an increasingly scientific aspect. Through artificial fertilization of eggs outside of the natural habitat, humans could produce fish on a scale that nature could not match. And, since fish could be hybridized, just as one can cross different versions of daffodils or iris, fish culturists could develop new varieties of fish with desirable characteristics, such as weight gain or temperament. Because fish culture will play a central role in our story, to understand how fish culture came to America we must first look to Germany and France.

FISH CULTURE IN EUROPE

Sometime during the mid-1700s, a German agriculturist named Stephen Ludwig Jacobi discovered—well, he probably learned it from local fishermen—the process of how to artificially fertilize fish eggs, specifically salmon and trout. Jacobi gets credit because he first described the process in a 1763 publication of the *Magazine of Hanover*. Basically, it amounts to this: harvest eggs (roe) from a female fish and milt (sperm) from a male fish, then mix the two together to fertilize the eggs. The fertilized eggs are then incubated in "hatching boxes," where cold water is run over them until they "hatch." (For future reference, remember that this was 1763.) The recently hatched fish—called "fry"—are transferred to rearing ponds and raised until a desired size is achieved. This may sound amazingly simple, but successfully accomplishing the outcome isn't: incorrectly mixing the eggs and milt may not yield enough fertilized eggs; the water in which the eggs are placed may not be the correct temperature and, therefore, the eggs would fail to develop; disease might derail the process; and predation of the young in the rearing ponds all could spell doom for the fry. Even with these potential complications, the success of Jacobi's experiments demonstrated that humans could intervene in the reproduction of salmonids. Perhaps we could lessen our dependence on nature to provide sufficient quantities of fish to meet our dietary needs?

Jacobi's results so impressed King George II of England—a German by birth—that the king awarded him a life pension. Jacobi's peers were not so impressed, however. The Royal Prussian Society of Sciences and the Northern Society of Sciences at Hesse gave Jacobi's discovery little attention. Even so, once Jacobi's writings were translated into Latin, French, and English, his process and the results of his experiments found a much wider, more accepting audience.

A French fisherman named Joseph Remy also experimented with controlled reproduction of fish.[27] Being illiterate, Remy could not have read Jacobi's publications. The provenance of Remy's discovery is unknown, so let's suggest that he "independently" discovered Jacobi's process to propagate trout. The story goes like this. Remy and his trusted friend and fellow fisherman Antoine Géhin had been experimenting for some years with the artificial propagation of trout on the La Bresse River in France. In 1843 (note the date) Remy sent a letter to the prefect of the Vosges describing *his* process for extracting eggs from female trout, milt from male trout to fertilize the eggs, and hatching large quantities of trout, which he raised to some size. Yes, it is the same process used by Jacobi, though he is not mentioned.

Remy not only had been able to fertilize and raise trout in a hatchery setting, but he also could produce so many that he even returned many of them back into the La Bresse River. Perhaps this is what Remy meant when he stated in his letter that his work might be of some interest to the government? It must have impressed the prefect and others: Remy and Géhin were awarded bronze medals from the Sociétè d'Émulation des Vosges. They also received a one-hundred-franc prize for their work. On top of these awards, they were rewarded with a state-licensed tobacco shop and life pensions from the state. Géhin even received an allowance to travel around France to promote the propagation process he and Remy had "discovered."[28] All of this public recognition encouraged others to try fish culture as a complement to other forms of animal husbandry. Remy and Géhin's work and the state's promotion of it put into motion a process that, if done on a large enough scale, could produce sufficient numbers of fish as a marketable commodity as well as enough to restock waterways where fish populations were declining because of overfishing. It was an early attempt to conserve a dwindling natural resource: fish.

The French government thought so highly of fish culture's possibilities that in 1852 it constructed a *piscifactoire*—a fish factory—on eighty acres near Huningue, in Alsace. Complete with hatching houses and rearing ponds, the operation represented the high point of modern fish culture. At Huningue the government produced millions of fish each year for consumption and for stocking streams. It is estimated that in less than a decade the facility produced more than one hundred million fertilized trout and salmon eggs, many of which were distributed throughout France. Some were even exported to England and nearly a dozen other countries.[29]

The notion that "solving" the fish problem could only happen if the government became involved is what made the operation at Huningue so important. The work at Huningue provided a template: if the government would fund the mass production of fish using artificial propagation, then it could "solve" the problem of declining fish populations. Science would prevail where nature failed. Now "aquatic resources [fish] would no longer simply be harvested; they would be manufactured and reinvented so that by kind and abundance they would be superior to the raw materials provided by nature."[30]

The revolutionary ideas underlying the operations at Huningue became the cutting edge of fish culture. In 1853 Victor Coste, the director of the fish factory, published *Instructions Practiques sur la Pisciculture*, one of the first how-to books from which anybody could learn and apply the latest techniques of fish culture. Coste's book, which popularized the techniques used at Huningue, was so important that it was translated into several different languages, including English. Coste's work would spark the introduction of fish culture in America.

FISH CULTURE COMES TO AMERICA

I want to focus on a few of the individuals who were at the forefront of this new field that combined mass production and science. Some of these men will have significant influence on the U.S. government's program to manage the nation's fish stock during the latter half of the nineteenth century. And if you think that the story of fish culture in America begins in some Eastern state, you'd be wrong. To see how fish culture came to the states we must visit the outskirts of Cleveland, Ohio.

Theodatus Garlick, a practicing physician in Cleveland, had read a translation of Coste's book. Intrigued by this new science and its commercial possibilities, Garlick and his friend H. A. Ackley, also a doctor, set about to try their hand at propagating eastern brook trout. Late in the summer of 1853, the two doctors acquired adult brook trout from two sources, one in Sault St. Marie, Michigan, and one in Port Stanley, Ontario. The trout were placed in ponds on Ackley's farm outside of town. Once spawning behavior was observed, eggs were harvested from the females and mixed with milt taken from the males. After incubating the eggs in running water, as per Coste's instructions, the eggs hatched after several weeks. Although there were some difficulties, their first attempt produced a batch of surviving fry. Encouraged by their initial success, the budding fish culturists made additional attempts and again were successful—they were so successful, in fact, that soon a successful fish-production enterprise began. The news of these two amateur fish culturists spread. Their enterprise became widely known, so much so that they were invited to display trout resulting from their activities at the Ohio State Fair. Their novel achievements even led to several awards. Notably, they were rewarded and praised not for the scientific aspect of what they had done, but the *practicality* of their accomplishment.

Garlick wrote up the results of his work with Ackley in a series of articles that appeared in *Ohio Farmer*, a weekly farm newspaper out of Cleveland. The intent was to demonstrate to those who owned a bit of land and suitable body of water that they too could raise fish for food and profit. These articles formed the basis of Garlick's book *A Treatise on the Propagation of Certain Kinds of Fish*, which he published in 1857. Garlick gives scant credit to the earlier work done in France. Rather, he adopts a definite U.S.-centric tone: "The work [in France] is valuable for the reason that it gives a detailed history of the progress that Fish Culture has made in Europe; besides much information that is valuable in a practical point of view. I am of the opinion, however, that whoever reads it, will agree with me, that it is deficient in some important points," the most essential being that "with the exception of the *Salmo Solar*, the habits of not a single American fish are given."[31] Like the early U.S. railroads that "adopted" English railroad technology, Garlick basically lifted the techniques that already were being practiced in France and claimed ownership. Garlick so depended on existing French fish culture that his book even includes crude reproductions of illustrations originally found in Coste's work. And, when mentioning him, Garlick doesn't even bother to spell the French pioneer's name correctly. While one may quibble over his cavalier use of another's work, Garlick's book quickly became the primer for up-and-coming fish culturists in America. Indeed, his early contributions were thought so influential that the American Fish Culturists Association (which I will discuss later) at their March 1881 annual meeting formally recognized Garlick as America's first fish culturist.

Even though Garlick got the designation of America's *first* fish culturist, the loftier label of "Father of American Fish Culture" has been given to another early practitioner, Seth Green.[32] Green was born in Genesee County, New York, in early 1817. Early in Green's life the family moved to the town of Carthage, a few miles south of Rochester. Here Green attended school, but only through the sixth grade. Even at a young age Green was a keen outdoorsman. He became so adept at catching fish that he would make it his livelihood.

Green opened his own fish market in 1848. He became increasingly interested in fish culture and in 1864 paid $2,000 (about $36,000 today) for a mill site near Caledonia, about seventeen miles outside of Rochester, where he built his own "fish factory." Green dug rearing ponds, filling them with water diverted from a spring on nearby Caledonia Creek. With some guidance from another fish culturist named Stephen Ainsworth, Green's hatchery became an immediate success. Green sold one-half interest to A. S. Collins in 1865 for $6,000, equivalent to about $116,000 today. Together they established a very successful business. By 1868 profits from the business had jumped to $10,000 (over $200,000 in modern terms). Though commercially successful, Green's interest in fish culture turned from merely producing fish for market to the "scientific" side of what this new field of experimentation could accomplish.

Green's business success arose from two sources. One, he was proficient at producing and selling fish for the market: brook trout were selling for about a dollar per pound, equivalent to a day's average wage in those days. Second, Green sold fish eggs to other fish culturists. Livingston Stone, a contemporary of Green's that you will meet shortly, explains why Green became so successful and renowned among his peers:

> Amateur and scientific experiments on a small scale had been made by various persons at various times, and the method of hatching fish artificially had been known for a century, but it remained for Seth Green to introduce into America the hatching of fish as a practical and valuable industry, and to him belong the credit and the honor of opening the way to the vast practical work that has since been accomplished in this country in hatching and rearing fish.[33]

Green's work and his discoveries in fish culture became renowned. Early on "the story about the man who was hatching thousands upon thousands of trout steadily gained ground. Presently the great New York dailies took it up, and soon after it came to be an accepted fact that something very wonderful was certainly being done by this New York trout-hatcher."[34] Stories written by him and about his work appeared in popular media outlets, such as local and major national newspapers, and in national outdoors publications like *Forest and Stream*, a major outlet for those interested in the latest developments in the growing fish culturist movement.

Green's fame also grew because of his experiments in cross-breeding species of fish in an attempt to create—much like gardeners hybridize day lilies or roses—a fish that would grow faster, exhibit less aggressive behavior in hatchery settings, or possess the genetic traits that allowed it to better withstand the rigors of nature if released into rivers and streams. Believe it or not, at the time his seemingly arcane accomplishments were being followed by the public.

Even the *Clinton Advocate,* a newspaper from a small western Missouri town, reported on Green's work. In one article it quotes Green as saying:

> "We cross the female salmon trout with the male brook trout and thus produce a hybrid. Then we cross the hybrid with the brook trout, which gives us three-quarter brook trout and one quarter salmon trout." Why do all of this? Because, says Green, the resulting fish "has all the habits of the brook trout, lives in both streams and lakes ... rises readily to a fly, is far more vigorous and fully one-third larger than ordinary brook trout of the same age."[35]

Perhaps one of Green's most valuable contributions to the field of fish culture in America involves his experiments with shad. Stone recounts that

> while so many [fish culturists] at first went to raising trout, no one seemed to even think that it was worth while to hatch any other kind of fish; and it is also a fact worth noticing that if artificial fish-culture had been confined to the raising of trout, as it was the first three years of its career in this country, the vast and beneficent work that is being done at present would have been unknown. It again remained for the bold and adventurous spirit of Seth Green, with his far-reaching vision, to enter the larger and more important field of hatching fish that had a standard commercial value. Everyone knows of his attempts, his failures, and his final success in hatching shad. These efforts of Green, in demonstrating that other and more valuable fish could be hatched as easily as trout, did indeed open up a field for fish-culture so vast and beneficial to mankind that the previous trout cultural work shrank into insignificance beside it.[36]

Green's publication of his how-to book *Trout Culture* in 1870 further expanded his reputation and influence, not only here in America but also abroad.[37] Green's importance to the field of fish culture is exemplified by his many national and international awards, including gold medals in 1872 and 1875 from the Impériale d'Acclimatation of France, a certificate of award from the U.S. Centennial Commission at the International Exhibition in Philadelphia in 1876, and a gold medal from the German Fishing Society in 1880. Green's pioneering work was instrumental in shifting the epicenter of fish culture from Europe to America.

Another notable character in this brief biography is the aforementioned Robert Barnwell Roosevelt. He will play a major role in the introduction of fish culture to the United States, especially at the national level. Roosevelt and

Green, despite their different social backgrounds, became close friends. This was likely because of their mutual interests. They enjoyed each other's company on numerous fishing outings over the years and, perhaps more importantly, shared similar views about the usefulness of fish culture to stem the decline in fish populations. While both were serving on the State of New York's first fish commission they even coauthored the book *Fish Hatching and Fish Catching* (1879), which they used to explain how fish culture worked and to advocate for fish culture as "absolutely necessary in order to the preservation of the fish of the country from total destruction." Echoing what the French had accomplished earlier, they pressed the idea that "individual enterprise is sufficient for success, though State action is desirable."[38] Roosevelt's political connections and dedication to the government's use of fish culture to replenish fish populations combined with Green's expertise and reputation made them a formidable duo in influencing what would become federal fish policy.

Lastly, the other figure in early U.S. fish culture that I will call out is Livingston Stone. Though Stone did not achieve the same public stature as Seth Green or Roosevelt, he significantly influenced the development of fish culture in America and the government's use of it. Like Roosevelt, Stone entered the world firmly ensconced into the upper crust of society.[39] Born in Cambridge, Massachusetts, in October 1836, his family could trace their paternal ancestors to early settlers of Plymouth Colony. His mother's side claimed direct kinship to the Winships, one of New England's most established families. Stone followed a path not uncommon to similarly positioned young gentlemen of his time. After graduating from Harvard in 1857, he entered the Meadville Theological School. Once his theological studies were completed, in 1864 Stone became the pastor of the Unitarian Church in Charlestown, New Hampshire. A career in religion apparently was not his calling, however. After only two years of tending his flock he left his pastoral position, purchased land near Charlestown and built a large fish hatchery which he named Cold Springs Trout Ponds.

Stone originally entered the fish business as just that, a business to produce trout for the market. But Stone ran his hatchery more like Green did his. He experimented, notably trying to propagate Atlantic salmon, which, heretofore, had not been successfully done at any commercial scale. Through this undertaking at Cold Springs Trout Ponds, Stone would achieve notoriety among his fellow fish culturists. In fact, he became one of the first to successfully raise Atlantic salmon on a scale not achieved in the United States.[40]

Stone's knack for raising Atlantic salmon in hatchery conditions achieved two important goals. First, it reinforced the idea that fish, even persnickety salmon, could be propagated and raised as one does other livestock to satisfy the dietary requirements of a growing population. Second, Stone's work suggested an even more exciting possibility: hatchery-propagated Atlantic salmon could be released back into northeastern streams and rivers to rebuild the dwindling population of native salmon. Was fish culture the answer to the perplexing problem of a dwindling salmon population?

Stone's experiments were made unduly expensive because he relied on Canadian sources of Atlantic salmon eggs. Canadian eggs were expensive: about forty dollars (equivalent to nine hundred dollars today) per 1,000 eggs. Because only a small percentage of the eggs survived to adulthood, experiments with Canadian-sourced salmon were not cost effective. Into the breach the recently established state fish commissions of New Hampshire and Massachusetts stepped. They agreed to jointly purchase fertilized salmon eggs from suppliers in New Brunswick with Stone acting as their representative in the negotiations. Under the auspice of the two state governments, Stone was given the task (and funds) to build and manage a salmon hatchery on the Miramichi River. Stone and his associates were able to obtain 440,000 salmon eggs in the first fall run, a majority of which would later become property of the Canadian government as per their agreement. After selling 100,000 eggs to his two benefactors, Stone kept the remaining eggs. Unfortunately, this initial venture into a government-backed attempt to propagate Atlantic salmon was abandoned after encountering unforeseen difficulties with the local authorities.

Unable to meet its initial goal, Stone's experimentation still produced some positive outcomes. Stone's involvement in the salmon hatching venture and his ability to transport salmon eggs from Canada back to his Cold Springs hatchery without great loss gave him a head start on others who also were attempting to propagate Atlantic salmon.[41] His experience and notoriety also positioned him to assume a key role in what would soon become the U.S. government's grand experiment to restore Atlantic salmon populations by transplanting Pacific salmon thousands of miles across country. The pivot to Pacific salmon was, of course, only possible because of one crucial fact: transcontinental rail travel recently had become a reality.

Like others Stone published a how-to fish culture book. He began working on it in 1870, though *Domesticated Trout: How to Breed and Grow Them* would not be published until 1873. The delay reflects the fact that during the interim he was working for the U.S. Fish Commission. Unlike other how-to fish culture books, Stone's includes important and detailed information on fish diseases and their treatments, the importance of water quality, and, significantly, how to successfully transport fish and fish eggs long distances. In the appendix "Journeys of Live Fish and Eggs," Stone recounts his attempts to ship fish and eggs. This discussion is meant to be instructive: here is what I tried, here is what worked, and here is what failed. For example, Stone tells the reader that in 1868 two lots of trout fry were sent to Providence, Rhode Island by express train without an attendant: all died. The lesson learned? "It is not safe to send live fish without an attendant, at least part of the way." In another instance a shipment of salmon fry was sent to the South Side Sportsmen's Club on Long Island, this time accompanied by an attendant. While en route water from the boat was used to change out the water in the shipping cans. The result was that almost two-thirds of the fry died. Stone's advice: "It is much safer to keep the fish in water that you are acquainted with than to use that with which you are not acquainted."[42] Stone also writes of long-distance shipments of fish and eggs that involved combinations of transportation methods—sled, cart, railroad, and even steamer. Stone's experience led him to believe that shipping eggs or fry any great distance is simply not possible without the use of railroads.

What these entries show is that Stone had built up a serious résumé of experience shipping salmon and trout. Unlike any other mode, Stone favored railroads for a faster and safer movement of fish and eggs. Stone's experiences will prove invaluable when he oversees the government's experiments to transplant fish from one side of the country to the other.

I have focused on these few individuals, but others also left significant marks on the early advances of the fish culture movement. Let me mention a few. Stephen Ainsworth of West Bloomfield, New York, began hatching trout as a hobby in 1859 and became proficient enough to produce enough fish to sell and to stock in nearby streams. He deserves mention because, as noted earlier, he provided a young Seth Green with advice and assistance early in the latter's career. Fred Mather began his fish hatching business in 1868 on a farm near Honeoye Falls, New York, close to Seth Green's facility. Mather will become a leading figure in the U.S. Fish Commission. Nelson W. Clark of Clarkson, Michigan, opened his first hatchery in 1868, then moved to Northville, Michigan, in 1874. From his Northville hatchery, the Commission will distribute thousands of trout and Pacific salmon to locations throughout the middle section of the country. No doubt there are others who probably deserve mention, and some of them will appear later. Some are important because of significant innovations in the fish production process, others for how they managed the government's experiment. For those individuals, I leave the telling of their histories to others.[43]

Governments Take Notice

A few states had begun by the 1850s to act against the worrying decline in fish stocks. Massachusetts established the first state fish commission in 1856. It lasted one year but was reappointed in 1865. That first year was productive, however. It recognized that fish culture probably offered the best chance to mitigate the loss of fish and so commissioned a report to summarize the state of European fish culture and provide an assessment of the Massachusetts fishery. In an appendix to that report, there is an account by one commissioner, Captain N. E. Atwood, of his attempts to propagate brook trout. Atwood's attempt failed miserably: all eggs collected died. Even so, it represented another early attempt at fish culture to solve the fish problem. It also added to the accumulating evidence that perhaps states, not just individuals, should use fish culture to rebuild fish populations within their borders.

Other state fish commissions soon appeared. From 1865 through 1870, nine state governments established fish commissions, including, by year:

   1865   Massachusetts, New Hampshire, Vermont;
   1866   Connecticut, Pennsylvania;
   1867   Maine;
   1868   New York;
   1870   New Jersey, Rhode Island.

During the next decade, eighteen more states formed commissions. Indeed, by 1879 roughly three-quarters of the existing states in the country had their own fish commission. The states that added a fish commission to their governmental structure during the 1870s are, by year:

1871	Alabama, California;
1873	Ohio, Wisconsin, Michigan;
1874	Iowa;
1875	Minnesota, Virginia;
1876	Kentucky;
1877	Colorado, Kansas, Nevada, West Virginia;
1878	Tennessee, Utah;
1879	Nebraska, South Carolina, Texas, Wyoming.

To get a feel for the obstacles the early commissions faced, consider the following excerpt taken from the 1867 biennial report of the Vermont fish commissioners. To set the picture, the report noted that Seth Green, under contract from the state, already had successfully propagated and stocked shad fry in the Connecticut River. The report provided detailed descriptions of personal visits by commissioners with notable fish culturists in England, Scotland, and Europe. These reports highlighted the attempts in those countries at reintroducing salmon to the local rivers. Even with this supportive evidence, the Vermont legislature would not vote to provide sufficient funding for such an undertaking. To this rejection the commission noted that

> there are many in New England who are uninformed upon the subject, and regard the whole theory of fish culture as a great humbug.... We are pleased to know that the more intelligent portion of the people of Vermont are not reckoned in this class, but feel mortified at the penuriousness manifested by those legislators who insisted upon limiting the entire expenses of five years to the paltry sum of five hundred dollars. But we are not disposed to harshly criticize the acts of these men, for they did not comprehend the importance or magnitude of the undertaking. Personally we care little about it, for we are so sanguine of success, that we propose to go on and do what we can to restock our streams, pay the bills, and risk the result.... While Massachusetts was expending her tens of thousands of dollars in constructing fishways, so that salmon and shad could reach New Hampshire and Vermont, it looked as though the people of [Vermont] were hardly willing to do the small share allotted to them, of merely furnishing the spawn for the young salmon and shad.[44]

Not direct enough? The commission's report ends with the observation that "many of the leading men of our sister States have turned their attention to the subject of restocking our streams with shad and salmon and are confident that under proper management it can be made a success."[45]

Fish commissions in those states whose waters emptied into the Atlantic Ocean often made restocking fish that made spawning runs—Atlantic salmon and shad—a priority. In other states the commissions adopted a more generic mission: stock fish, any fish, to improve the state's fishery. The Michigan legislature, for example, charged Michigan's Board of Fish Commissioners, established in 1873, to use fish culture "to increase the product of the fisheries." When the territory of Wyoming established its Board of Fish Commissioners a decade later, in 1882, they were similarly called upon to "in the most economical and practical manner, procure and distribute fish in the public waters as shall in their judgment best promote the increase and preservation of food fish."[46] Similar objectives are found in descriptions of most other state fish commissions. In fact, one analysis of reasons cited for forming state fish commissions found that 80 percent of the commissions had replenishing depleted fisheries and overharvesting as the main reasons given. Stopping the use of illegal fishing methods (dynamiting rivers and lakes was a popular method) came in a distant second.[47]

Why all this talk about fish commissions in a chapter about fish culture? Two reasons. First, the practice of fish culture by private individuals had spread significantly in just a couple of decades. Within a short time, there were more than one hundred known, active fish culturists across nineteen states. These were not mere hobbyists but individuals for whom fish culture was a commercial endeavor. Some, like Green, were actively experimenting with hybridizing different fish species, thus expanding the knowledge base of the field. Second, the expanding list of state fish commissions and the reasons given for creating them reveals that politicians (and the public) were starting to recognize the role that fish culture could play in replenishing fish stocks and making fish a reliable food source.

There also was a growing belief that only with the involvement of the government could the loss of fish stocks be adequately countered. The breadth of the problem made it unlikely that private fish culturists could produce enough fish for market consumption *and* to restock the nation's rivers and lakes. This push for governments to scale up fish propagation activities meant that an effective means to transport the necessarily massive numbers of fish eggs, fry, and adult fish across the country also was needed. And that meant any successful program, whether it was at the national or even state level, would rely heavily on the country's railroads.

The American Fish Culturists' Association

I end this chapter with an advance in U.S. fish culture that has nothing to do with its actual practice. Rather, it has to do with creating a coalition of practitioners. It has been suggested that such organizing was a move to exert more control over prices of eggs, fry, and fish, but that remains a topic of debate.[48] The fact is that in the summer of 1870 several well-known hatchery owners began contacting each other to discuss the possibility of meeting to create an industry association. Those initiating the meeting were Fred Mathers and J. F. Slack from Bloomsbury, New Jersey; A. S. Collins (business partner to Seth Green); Reverend W. Clift from Connecticut; and Livingston Stone. The details of this correspondence did not survive, but in early November 1870 an announcement appeared in New York City newspapers announcing an organizational meeting of "practical fish culturists." It was held on 20 December in the offices of the New York Poultry Society. In attendance was some of the country's best known fish culturists. In addition to the aforementioned gentlemen were Dr. M. C. Edmonds of the Vermont Fish Commission, Dr. J. D. Huntington of Watertown, New York, and B. F. Bowles of Springfield, Massachusetts. The august list of attendees indicates that those in the fish culture fraternity must have thought this would be an auspicious occasion.

If some of the conveners thought they could produce an agreement to fix prices on eggs and fry, there is nothing in the record to suggest that anything of the sort occurred. The most important outcome of this inaugural meeting was, however, an agreement to form an association and draw up a constitution. Indeed, the first article of the constitution states that "the name of this society shall be 'The American Fish Culturists' Association.'" The constitution also provides a clear mission statement: "to promote the cause of fish culture; to gather and diffuse information bearing upon its practical success; the interchange of friendly feeling and intercourse among the members of the association; the uniting and encouraging of the individual interests of fish culturists."[49] Annual membership dues were set at three dollars (about sixty-eight dollars in modern terms) and officers of the fledgling organization were chosen. Those in attendance elected Reverend William Clift of Connecticut as president; B. F. Bowles agreed to serve as treasurer, and Livingston Stone became the association's secretary. Deciding when the next annual meeting would be held was the only other big decision to come from the meeting. The next meeting of the association would be held in February 1872 in Albany.

Those in attendance agreed that the association should act as a forum to exchange new ideas about the propagation of salmon, shad, and trout. At its annual meeting in 1872 papers about recent findings in several areas of fish culture were read and discussed. Livingston Stone talked about issues confronted when trying to raise trout; William Clift spoke similarly about propagating shad; and Dr. M. C. Edmonds read a paper on the introduction of salmon into American waters. Given the earlier discussion of the efforts by the French (and later Germans), the following assertion of Edmond's is rather illustrative of the American attitude:

> Although England, France and Germany have done so much, yet it redounds not to the ultimate good of the people, but to the glory of individual enterprise, and the accomplishment of the object with them is the realization of large incomes to individual effort. The American idea seems to be utterly devoid of selfish consideration, being as it is for all the people, and for their continued prosperity. I conceive of no higher ambition for any man or set of men than the ultimate restocking our streams with the migratory sea fish, more especially the salmon. It at once gives all classes the advantages of cheap and desirable food. And, gentlemen, are we not commanded "to feed the hungry," and how better can this great duty be performed than by laboring to restock our lakes and rivers with fish of all kinds? To this end let us labor and eventually perpetuate a blessing.[50]

These were lofty, societal goals. These fish culturists were not merely owners of fish hatcheries but viewed themselves as the only solution to stem the decline in the nation's fishery. Theirs was an act of near religious zeal. Theirs was a noble goal. Interestingly, another more parochial outcome of the meeting was deciding how the association could more actively lobby governments at all levels for legislation to protect their industry. The now organized group of fish culturists set for themselves a path to elicit government protection and funding for their activities. This is clear in A. S. Collins's motion to "recommend that the Legislatures of the different States pass such laws as shall encourage and protect pioneers in fish culture."[51] Exactly what form of encouragement and protection is not spelled out. Even so, the motion passed handily. A noble goal, indeed.

Another motion passed at this meeting would have more lasting effects. George Shephard Page, an ardent trout enthusiast and, more importantly, a member of the New York Fish Commission, moved to establish a committee to draft and submit to Congress a "memorial" calling for the federal government to fund two fish propagation operations.[52] One would be located near Puget Sound on the West Coast and would focus on propagating Pacific salmon. The other would be located at "some convenient point near the Atlantic" and concentrate on the propagation of shad. In both instances the aim was to propagate salmon and shad in such large numbers that they could be used to restock eastern rivers and streams. Page's suggestion and the association's unanimous approval of it are critical in the history of fish culture in America. Henceforth, the association's members actively lobbied the federal government (and state governments) and became increasingly entwined with the government's attempts at managing the nation's fish stock.

One aspect of this meeting signals just how entangled the association would become with the government's program to manage the fishery. At the meeting the members bestowed upon Spencer Fullerton Baird the title of honorary member. A well-known naturalist in his own right, Baird was not a practicing fish culturist: he did not even own a hatchery. So why did he rate such an honor? Simple: Baird recently had been appointed commissioner of a new government agency, the U.S. Fish Commission. It should not be too surprising, therefore, to find that many of the individuals mentioned as founders or early members of the association will become employees of and advisors to the U.S. Fish Commission in its attempts to manage the nation's fishery.

A Final Note

The "modern" approach to fish culture began in the 1700s in what is today Germany and later in France. Early practitioners in America borrowed heavily from those early pioneers, adapting (often claiming for themselves) what others had done as befit their own individual needs and situations. Starting in the 1850s, the success of fish culturists improved with the continued introduction of new techniques and the discovery of solutions to problems that plagued the practice. These advances and finding better ways to transport fish and eggs strengthened the belief that fish culture might just be the answer to the problem of a depleted fishery.[53] Before too long the propagation of fish in hatchery settings for return to the wild was believed to be the only viable way to solve the nation's ongoing fishery crisis. Indeed, it is arguable that the evolution of fish culture, from the earliest attempts in Europe to the developments in the United States, represents the beginnings of a conservation movement that continued throughout the 1800s and is still in effect to this day. From the late 1860s onward, the advancing technology of fish culture and the expanding horizons of the railroad will merge with significant consequences, not only for the government and its attempts at fish management, but also for the development of sport fishing across the country.

3

THE U.S. FISH COMMISSION

Up to now I've considered two technologies that emerged in the first half of the nineteenth century: railroads and fish culture. The relatively small fraternity of fish culturists already used railroads to transport fish eggs and fry in the 1850s and 1860s. There just weren't that many fish hatcheries that had enough business to be transporting that many fish long distance by rail. That would change dramatically during the remainder of the century. The reason? The completion of the transcontinental railroad and the formation of the U.S. Fish Commission. Created by an act of Congress in 1871, the U.S. Fish Commission—hereafter, the Commission—would depend on the railroad industry to carry out its policy objectives. To understand what drove Commission policies through the end of the century, it is informative to first explore the squabble that started it all.

Line Fishermen versus Trappers

You've already seen that by the late 1860s the poor condition of the nation's fishery was widely known. In an 1868 article, *Harper's New Monthly Magazine* provided a lengthy list of causes for the decline in fish stocks. These included the usual suspects: too many dams built on rivers and streams that once supported spawning runs; city sewer systems emptying waste (human and industrial) into convenient streams; and, of course, overfishing. A decade earlier Marsh had made a similar argument in his report to the Vermont legislature. But the *Harper's* article represented something different: because the list appeared in the widely read *Harper's* magazine, it grabbed the public's attention.

A major issue concerned how the fish—and here I focus on sea-run fish like shad and salmon—were caught. New England fishermen used two methods to harvest sea-run fish: by line and with traps. Line fishermen operated from small boats or from the shore using a single line with a baited hook. Trappers, on the other hand, fixed long nets often placed at the mouths of rivers and captured whatever came through. As the fish entered the rivers to spawn upstream, trappers harvested fish in much greater numbers than could line fishermen. Trappers claimed, rightly, that only by using their methods could the market demand for fresh fish be satisfied. Their approach, they argued, kept market prices in check and made fish more available to the public. Intermediaries between the trappers and the consumers—not only locally but with improved rail transportation increasingly to inland cities as well—agreed. Consumers, many of whom considered fish an integral part of their diet, could care less about how their fish got to market as long as it remained available and reasonably priced. Line fishermen simply could not compete with the trappers. Watching their age-old livelihood disappear, they turned to their elected officials for protection.

The dispute became a political hot potato. In response to line fishermen claiming that trappers prevented shad from making their springtime spawning run up the Connecticut River, the Connecticut legislature passed legislation in 1868 that restricted the use of traps at the river's mouth. The actual enforcement of this law, like most other fishing regulations of the time, was unimpressive. In neighboring Massachusetts and Rhode Island, the dispute between trappers and line fishermen was similarly heated. These legislatures did not react the same as Connecticut, however. Massachusetts, basing their policy response on a report conducted by Nathaniel E. Atwood, a self-taught ichthyologist, favored the trappers and rejected the petitions of the line fishermen. In Rhode Island the state's constitution said that those fishing from the shore (line fishermen) enjoyed rights equal to those fishing in the sea (trappers). Most assumed the legislature would therefore find in favor of the line fishermen and regulate the use of traps, but that did not happen. After considerable debate, and noting that Massachusetts had failed to limit trapping, the Rhode Island legislature declined to act.

This is an interesting example of how governments assign (or don't) property rights where none naturally exist. But what has it got to do with the creation of the Commission? Everything. Spence Fullerton Baird, a nationally recognized naturalist and the assistant secretary of the Smithsonian Institution at the time, had developed an interest in marine biology. The line fishermen versus trappers debate and how it fit into the concern over observed fish declines aroused his scientific curiosity.

Spencer Fullerton Baird

Baird was born in Reading, Pennsylvania, in 1823.[1] His father, Samuel, was a prominent, politically active lawyer who served in the Pennsylvania Senate. Baird's mother, Lydia Biddle Baird, was descended from the Biddles of Philadelphia. Spencer grew up in, needless to say, a well-to-do household.

After his father's death in 1833, the family moved to Carlisle, Pennsylvania, where Baird's mother's relatives lived. In Carlisle Baird attended a grammar school associated with Dickinson College. As a young boy Baird exhibited a deep interest in the natural world around him and was an active and successful collector even in his early years. It shouldn't be surprising, then, that as an undergraduate he studied natural history at Dickinson, where he earned his degree in 1840. Baird initially sought a career in medicine, enrolling in the College of Physicians and Surgeons in New York City. While taking medical classes, however, he took every opportunity to continue his study of natural history. Because his passion lay not in medicine but in the study of the natural sciences, it was not surprising that he left medical school and returned to Dickinson. He continued his studies in natural history and earned a master's degree in 1843.

Baird, who was actively engaged in a variety of scientific studies, became acquainted with a number of eminent natural scientists of the day. He spent considerable time with the noted artist and naturalist James Audubon, who mentored the young Baird, improving his skills as a naturalist and shared a common passion: birds. His friendship with Audubon opened doors to a wider scientific community for Baird. Not surprisingly, when Dickinson offered him the position of professor of natural history in 1845, he took it.

Over the next five years Baird's activities revealed a side of his personality that would resurface later in life. He proved to be a conscientious and inquisitive young scientist, amassing a large collection of various animals. He was a popular professor and took on additional administrative duties, serving as chair of both the natural history and chemistry departments. In addition, Baird acted as curator of the college's natural history collection. His penchant for detail and organization would be great assets in his future endeavors.

During his time at Dickinson, Baird's fieldwork increased the size and scope of his personal collection. He amassed a wide array of birds, reptiles, and fish.

His ever-growing set of acquaintances even included the likes of the preeminent natural scientist of the day, Louis Agassiz of Harvard University. Agassiz made many contributions across several fields but is perhaps best known for his scientific approach of gathering and analyzing data. Baird honed his scientific skills, a hallmark of how he tackled issues faced later in his career.

Baird also demonstrated an aptitude at making the kind of contacts—professional and personal—that helped advance his career. This is not a pejorative comment: Baird was quite deft at using his connections among the group of eminent scientists and politicians to advance his causes. This skill would be important later in his career at the Commission when he cajoled Washington politicians to vote for his projects.

Baird's writings earned him a deserved reputation as one of the country's best naturalists. Before long opportunities outside of academia arose. Given his ambitious nature and his scientific skills, the confines of Dickinson College became a bit cramped. And so, in 1850, at the age of twenty-seven, Baird applied for the vacant position of assistant secretary of the Smithsonian Institute. With the support of several of his eminent friends—such as George Perkins Marsh, Agassiz, and others—Baird's appointment was quickly approved.

Baird's years at the Smithsonian have been dealt with elsewhere, so I will not detail them here.[2] Baird excelled at his job, balancing his scientific interests with his growing administrative duties. For example, he helped organize and expand the institute's collection of scientific books and manuscripts. He oversaw the publication of its scientific studies in a growing number of areas. He frequently wrote columns on science for popular publications, further expanding the reach of the Smithsonian—and of himself—to the general public. Baird's activities were helping to improve and promote the national and international status of the Smithsonian in the scientific community. He also became familiar with the ways of Washington politics, something that would become a great asset in his future career plans.

In addition to his administrative activities, Baird oversaw a substantial expansion in the Smithsonian's zoological collection. When he joined the staff, the Smithsonian's natural history collection was almost nonexistent. He substantially increased the collection by donating his own larger collection, which amounted to hundreds of glass jars, barrels, and other containers that held a variety of plants and animals. It is reported that it took two freight cars to haul it all from Carlisle to the institute in Washington DC. Baird wanted to make the Smithsonian the nation's repository of all things natural history. Because some of the best collections were in the hands of private institutions, Baird actively (and successfully) lobbied many of these individuals to donate their specimens to the Smithsonian.

With the nation's boundaries still expanding and much of the West yet to be explored, one of Baird's more ingenious ideas to build the institute's collection was to piggyback off the various railroad survey expeditions by the U.S. government to explore and map the western United States.[3] Baird recognized the potential for mapping not only the distribution of species, but also the ability to collect specimens of the many yet unknown species. Given this potential, Baird and the Smithsonian supplied the necessary scientific equipment and instructions used by collectors assigned to these survey parties. Of course, when the expeditions returned, the Smithsonian received all the collected items.

By the late 1860s, Baird, already an eminent scientist in his own right, decided that if the fish problem in New England was going to be solved, it needed some scientific analysis. And he was just the person to carry out that investigation.

Baird and the Dispute

As a frequent summertime visitor to Woods Hole, Massachusetts, Baird was aware of the complaints lodged against the trappers. But were these complaints scientifically sound? That is, even though the trappers harvested more fish, were they the primary cause for the decline in spawning runs? Baird the scientist believed that before any sensible regulatory decision could be made, the problem should be studied scientifically: hypotheses posed and tested. By "test" Baird meant that he would gather as much information as possible before making a verdict. This involved collecting harvest numbers over time, talking with the parties involved, doing field work, and so on. Thus, when Baird returned to Washington from Woods Hole in the fall of 1870, he decided to seek a government appropriation for the study of the coastal fishery problem.

Baird thought $5,000 (equivalent to about $114,000 dollars today) would be sufficient to undertake his study that, if done properly, would take a few years. Familiar with the ways of Washington, Baird went directly to the source of governmental funding: Henry L. Dawes. Dawes was the perfect target for Baird's lobbying. He served as congressman from Massachusetts, a state directly involved in the dispute and for whom the industry of coastal fishing was important to its economy. Perhaps equally important, Dawes was the powerful chair of the House Appropriations Committee.[4] Dawes found Baird's proposed study interesting and worthy of governmental funding. Wouldn't Dawes want to be known back home as the one who helped make such a potentially important study possible? But there was a snag: Baird's funding could not be attained by just attaching some rider to a bill and waving it through the appropriations process. Funding Baird's project required that a whole new government agency would be needed.

The Resolution and the Commission

Baird wrote to Dawes in January 1871 formally requesting the formation of a federal fish commission whose sole purpose was to conduct a study of the decline of coastal fishes. Armed with a supporting letter from the well-known and eminent Smithsonian scientist to support his request, on 18 January 1871 Dawes introduced a resolution before Congress. The resolution proposed to create a federal commission whose purpose would be to "protect and preserve the food fishes" of the coastal United States.

Dawes's resolution met with skepticism, the kind that has long characterized political debate in this country—that is, East Coast interests versus the rest of the country. Was the honorable Mr. Dawes of Massachusetts suggesting that the problems of coastal fishes were more important than the problems faced by fisheries in interior states? During discussion of the proposal, Representative John F. Farnsworth of Illinois even suggested that the problem of coastal fishes should be of no greater importance than that of potato bugs, presumably a problem affecting his constituents. After the debate, when it came to a vote the resolution failed.

This setback did not deter Baird, however. Through his years at the Smithsonian, he knew how and whom to lobby to get his project funded. Baird solicited help from other Massachusetts congressmen who supported the resolution. He met with representatives from Indiana and his home state of Pennsylvania. He even met personally with Congressman Farnsworth, to offer him a greater perspective of the problem. The decline in coastal fish, explained Baird, actually had national implications and did not just serve the parochial interests of the East. Demand from a growing population facing a declining supply of fish would cause prices for seafood and other types of fish to increase all across the country. If fish became scarce, people would have to substitute other sources of protein, which would drive their prices up. If all that came to pass, would Farnsworth want to be the one blamed for rising consumer prices and the resultant lack of protein in the average diet?

A savvy negotiator, while making his case for the resolution Baird also sought to discover just what it would take to gain a congressman's backing. With this information Baird modified the original version of the resolution to broaden its appeal. The resolution's original wording focused on a study of coastal fisheries. The revised proposal broadened this to include investigating problems with other fisheries, especially in Lake Michigan, where reports of falling whitefish populations were alarming the locals. Essentially, Baird broadened the proposed commission's investigations to encompass the coast and *all* lakes in the United States. This seemingly minor word change played to the regional interests of the politicians voting on it. Now politicians from various areas of the country could claim that *their* vote for the resolution reflected *their* requirement that activities of the new fish commission would benefit *their* constituents. Another change to the original resolution required that the fish commission be led by a current civil officer of the government—in other words, an existing government employee who would assume the added duties of managing this new organization. Oh, and for this the person would not receive any additional compensation. Who would be willing to undertake those additional responsibilities without additional pay?

All of the foregoing activity took place across a few days. The revised resolution was reintroduced to Congress on 22 January 1871. The *Congressional Globe* reported that the resolution passed, 127 yeas to 48 nays. Representative Farnsworth maintained his opposition. Missouri congressman John F. Benjamin joined the anti-commission bloc and presciently claimed that if passed "there will be no end to the expenditures of public money before we get through with [the study]."[5]

The resolution then moved to the Senate for debate. Once it passed the House, Baird immediately began to lobby senators. The day after the House vote he met with Senator George Franklin Edmunds of Vermont. Senator Edmunds was not only an influential politician, but also a prominent supporter of early conservation actions and the fish culture movement. Remember Dr. M. C. Edmunds, a member of the Vermont Fish Commission and a founding member of the American Fish Culturists' Association? He was

Senator George Franklin Edmunds's cousin. In addition, as I noted in chapter 2, Senator Edmunds was a personal friend of Baird—the two met over the years at Woods Hole. He was also a confidant of President Grant. Baird met with Zachariah Chandler of Michigan as well. Senator Chandler would be an important ally in this campaign, because Michigan would benefit from having a fish commission charged with studying the problems plaguing the Lake Michigan fishery. Chandler also happened to be the chairman of the Senate Commerce Committee.

Over the next few days Baird lobbied a number of other senators, including Roscoe Conkling of New York, Timothy Howe of Wisconsin, and Allen Thurman of Ohio. Baird was lining up votes from senators representing different regions of the country. He wanted support from those whose constituents, individuals and businesses alike, would benefit from the Commission's work on coastal *and* inland fisheries. Baird's lobbying efforts and the support of other influential backers paid off: the resolution moved quickly through the Senate. On 24 January the resolution was referred to Senator Chandler's Commerce Committee. Two days later it was reported out of committee, and on 6 February the resolution, along with other items being ushered through by Senator Chandler, passed without debate.[6] A few days later the resolution landed on President Grant's desk for his signature, which he affixed on 9 February 1871. With his signature President Grant created the position of U.S. commissioner of fish and fisheries. Little did the president know that in signing the resolution he set into motion one of the more encompassing and enduring experiments in applied fish culture yet attempted.

With the authorization for the Commission in place, the action returned to the House for a vote to secure the $5,000 that Baird originally requested. Once again Congressman Dawes helped by inserting the appropriation into a funding bill including various projects. The early opponents of the Commission seemed to have lost the will to fight it any further, and on 3 March 1871 the funding bill easily passed. Baird had his funding to support his research, and more.

I should not leave you with the impression that Baird ought to take all the credit for passage of the resolution and creation of the Commission. The difficulties with declining fish populations and the apparent inability of states to look beyond their own provincial benefits were creating an atmosphere in which reliance on the states or the private sector to solve the problem was fading. By 1870 a number of individuals were lobbying Congress for some form of federal action, even though there remained a latent distrust of federal government interference in what many still perceived as a state issue. This helps explain the debate over the question of whether the government should create a new department or, as it did, an agency with no regulatory oversight. Even though the decision took the latter route, history shows that, in the end, the Commission would become a prominent force in fish management and, under Baird's leadership, a political force as well.[7]

Now the only thing left to do was find someone who had the necessary background and would be willing to serve as the new U.S. fish commissioner. The resolution stated that the person must be someone "of proved scientific and practical acquaintance with the fishes of the coast" and who, as already noted, would be willing to "serve without additional salary." Given his stature among naturalists, fish culturists, and politicians, not to mention the fact that the resolution had Baird's fingerprints all over it, Baird was the frontrunner—indeed, the only contender—for the job. Enter once again Senator Edmunds of Vermont. He escorted Baird to interview for the job with President Grant the day after he had signed the resolution. It probably took little persuasion from his friend Senator Edmunds before the president formally offered the position of fish commissioner to Baird, which he accepted. The Senate formally approved Grant's nomination of Spencer Fullerton Baird to be the fish commissioner of the United States on 8 March 1871.

The Resolution

The official title of what I've been calling the resolution is *The Joint Resolution for the Protection and Preservation of the Food Fishes of the Coast of the United States*. Even though it would permanently set a course for the government's intervention into managing the nation's fishery, it is surprisingly short. Let's reproduce it here in its entirety.[8]

> Joint Resolution for the Protection and Preservation
> of the Food Fishes of the Coast of the United States.

Whereas it is asserted that the most valuable food fishes of the coast and the lakes of the United States are rapidly diminishing in number, to the public injury, and so as materially to affect the interests of trade and commerce: Therefore, Be it resolved by the Senate and House of

Representatives of the United States of America in Congress assembled, That the President be, and he hereby is, authorized and required to appoint, by and with the advice and consent of the Senate, from among the civil officers or employees of the government, one person of proved scientific and practical acquaintance with the fishes of the coast, to be commissioner of fish and fisheries, to serve without additional salary.

SEC.2. And be it further resolved, That it shall be the duty of said commissioner to prosecute investigations and inquiries on the subject, with the view of ascertaining whether any and what diminution in the number of food fishes of the coast and the lakes of the Unites States has taken place; and, if so, to what causes the same is due; and also whether any and what protective, prohibitory, or precautionary measures should be adopted in the premises; and to report upon the same to Congress.

SEC.3. And be it further resolved, That the heads of the executive departments be, and they are hereby, directed to cause to be rendered all necessary and practicable aid to the said commissioner in the prosecution of the investigations and inquiries aforesaid.

SEC.4. And be it further resolved, That it shall be lawful for said commissioner to take, or cause to be taken, at all times, in the waters of the sea-coast of the United States, where the tide ebbs and flows, and also in the waters of the lakes, such fish or specimens thereof as may in his judgment, from time to time, be needful or proper for the conduct of his duties as aforesaid, any law, custom, or usage of an State to the contrary notwithstanding.

APPROVED, February 9, 1871

The resolution from the outset makes clear what the problem is: the "rapidly diminishing" number of fish—more importantly, "food fishes." The bill was not meant to improve fishing for the sportsman but to remedy the loss of a protein source for the citizenry. As I noted earlier, fish were a staple in the diet of many households. Cities in the country's interior consumed increasing amounts of oceangoing fish, thanks to advances in shipping across the expanding railroad network. Demand incessantly pressed on the limited supply of salt- and freshwater fish. It was an old problem, but it was becoming worse. Some pointed to the difficulties in Europe, especially France, and warned that the United States would soon face the same fate without intervention. Additionally, the livelihoods of those in the fishing business—commercial fishermen and fish dealers—suffered with the decline in fish stocks.

Notice that even though the focus of the resolution is on coastal fishes—it is, after all, in the title—this constraint is relaxed in the text where it adds "and the lakes of the United States." Added to appease the congressmen and senators from the interior states, the insertion of these few words expanded the Commission's range of activity—and the funds to carry out such study. Indeed, Missouri congressman John F. Benjamin's quip about future funding is prescient.

Section 3 of the resolution is important because it requires heads of other government agencies to render assistance to the Commission. This will prove vital to the Commission's oceanic work as it made government ships available for oceanic experiments and explorations. Section 4 must have thrilled Baird the naturalist. It allows the Commission to collect any and all species in or near the waters that it (the Commission) deemed necessary to advance the scientific community's knowledge of the country's environment. In essence Section 4 allowed the Commission to engage in a governmentally funded program of acquisition that would greatly expand the Smithsonian's collections of flora and fauna. As an intimate of Baird's wrote, the Commission created by this resolution gave Baird "the opportunity of making immense collections, of which he was not slow to avail himself, and there was enough utilitarianism in the work, that could be appreciated by everyone, to make it secure ample appropriations from Congress for carrying it on."[9] Baird could not have asked for more.

Back to the Dispute

Baird wasted no time in taking advantage of the opportunities that came with being the newly minted U.S. fish commissioner. Granted a leave of absence from the Smithsonian, Baird arrived at Woods Hole in June 1871 and set up shop to begin his study. He knew Woods Hole from his times visiting, and it offered a convenient base from which to conduct his enquiries, which he made throughout the summer. The fruits of the summer's research are found in Baird's first report as commissioner that was officially submitted to the House and Senate on 31 January 1872. Because it wasn't printed until a year later, Baird took the opportunity to include the results of additional investigations made during the latter parts of 1872.

Baird's report and the accompanying materials contains a trove of information and data on the coastal fishery. His introductory comments run about thirty pages and set the tone for the report. Why conduct the study? What are the questions to be addressed? How did the joint resolution fit in? What were the limitations of his analysis? All the kinds of questions one would expect from a scientific inquiry. And, since there were limited inputs, Baird cautioned that any findings must be viewed as tentative, awaiting further analysis. The report also included a collection of papers and analyses by other investigators, and testimony given by interested parties at public hearings convened by Baird. It even includes results from an initial enquiry into the issues affecting the whitefish population in Lake Michigan. The report runs hundreds of pages, not counting over seventy-five pages of plates (drawings). All in all the report made quite an impressive first impression.

Remember that the objective of the study was to try and determine the cause of the decline in coastal fish populations. Much of the information collected, even including testimony of trap fishermen themselves, pointed strongly to the conclusion that, indeed, the increased use of traps, weirs, pounds, and the like over the years were a major cause of the observed decline in coastal food fishes. Even so Baird noted that these were not the only factors explaining the decline. Other sources were at work—for example, the oft-mentioned blocking of streams with dams that prevented fish from making spawning runs, industrial and urban pollution, lack of food, and climatic conditions. Baird did not, therefore, lay all the blame at the doorstep of trap fishermen. Instead, Baird hedged: "The conclusion appears warranted that if measures can be taken to prevent the present great destruction of spawning-fish, the supply will again increase before long.... At the same time, I am not prepared to advocate the abolition of traps and pounds, as without them it would be extremely difficult to furnish fish in sufficient quantity to meet the present and increasing demand of the country."[10]

Baird knew that he had not collected enough data to provide a firm, scientifically based conclusion. He also was astute enough to know that an immediate ban on trap fishing would severely reduce the harvest and create a 1870s version of a major supply chain disruption. With line fishermen simply unable to meet market demands, such a significant shortage of fish would drive the market price of fish upward in the coastal markets and in the emerging markets in western cities. If that were to happen, it would be even harder for households to put fish on the table. And that would also raise the ire of politicians, the very individuals that the Commission depended on for funding.

What was Baird's suggested solution? Based on the evidence collected—and, of course, more study would be necessary to reach any sound conclusion (translation: more years of study funded by the government)—Baird recommended that trapping be allowed to continue, but within limits: "I have come to the conclusion that if the capture of fish in traps and pounds be absolutely prohibited, under suitable penalties, from 6 o'clock on Friday night until 6 o'clock on Monday morning, even during a season of six weeks only, (thus required a close time of three nights and two days, to enable the fish [to] pass and perform their natural function of reproduction) the interest of all parties would be subserved."[11]

This made sense. If some spawning fish were allowed to make it upstream to do what nature was driving them to do, it would help ensure future generations and maintain the supply. Of course, the problem was not knowing just how many of these lucky fish it would take to rebuild the dwindling populations or just maintain a balance between harvest and natural reproduction. With the success ratio of spawning fish quite low, reducing the time when trappers could work probably would not do much to solve the problem. But it represented a middle ground between the two sides. Baird also invoked an economic argument to mollify the trap fishermen should his proposal be enacted:

> By permitting the catch as suggested, there is a greater certainty that the entire supply will be put to its legitimate use as food; and it is probable that, while less money may be made by the middlemen [who keep up the retail price] the owners of the pounds and traps would receive quite as large an amount of money for less labor and for three-fourths the same weight of fish. This arrangement would also furnish an opportunity for persons connected with the fisheries to repair their apparatus, or attend to other duties.[12]

Even though the scientific basis of Baird's proposal is questionable, it offered a politically expedient compromise to the feuding parties. And if the states involved should refuse or be unable to reach some sort of settlement along the lines he suggested, Baird saw no recourse but to impose federal action:

> In the event, however, of the refusal of the States mentioned [Connecticut, Massachusetts, New York, and Rhode Island] to establish the very limited close

time suggested, *I would recommend the passage by the United States of a law prohibiting, until further notice, the erection of fixed apparatus for taking fish, after a period of one or two years, on the south side of New England and on the shores of Long Island, which constitute the spawning-grounds of the shore-fishes referred to.* Although this would be a serious blow to the pound and trap interest.... The restoration of the fish to their original abundance would be thus accomplished in a much less time than by any merely palliative measures (emphasis added).[13]

In the end Baird's recommendations were for naught. Rhode Island seemed more amenable to adopting Baird's proposals, but once Massachusetts lawmakers were unwilling to agree, the chances for passage in Rhode Island were nil. Baird professed to believe in limited federal authority and preferred to let states make their own management decisions. Even so, this episode revealed the fact that states too often behaved with only a narrow viewpoint and were not likely to pursue conservation measures that created positive externalities shared by other states. Indeed, the outcome of this venture into conservation policy influenced Baird's views on how his commission would operate.

Baird "clearly failed to achieve the goals outlined in the Congressional resolution establishing his Commission. Persuasion and empty threats could not force a political settlement."[14] Baird's initial failure in his rookie year affected his behavior as commissioner in the future. He would adopt fish culture technology as *the* sensible solution to solving the fish problem, whether in coastal waters or inland lakes and streams. Under his leadership the Commission's portfolio quickly expanded to include nearly every aspect of the nation's fishery. His infant Commission would soon expand its operations in its attempt to manage the nation's fish stock.

The Commission Pivots

Recall that Vermont fish commissioner M. C. Edmunds, a prominent member of the American Fish Culturists' Association advocated for transplanting Pacific salmon into eastern streams to offset the loss of Atlantic salmon. Such a scheme would be too costly for any one state to fund, however. And, given recent experience, it seemed highly unlikely that several states would agree to pool their resources for this purpose. Hence, in his December 1871 letter to his cousin, Senator George F. Edmunds, Edmunds proposed that the federal government fund the building of a hatchery on the West Coast and one on the

East Coast. The idea was that fish would be propagated at the two hatcheries, and their fertilized eggs would be shipped to requesting state fish commissions (or anyone else), who would hatch them and stock the resulting fry into their streams. Edmunds's idea, backed by the American Fish Culturists' Association, was that the federal government should subsidize states' efforts to offset the loss of fish that had significant commercial and even sporting value. Edmunds's request was similar to a proposal that had been suggested earlier to Baird by Robert Roosevelt and Seth Green.[15] Baird received a related idea from William Clift, another founding member of the American Fish Culturists' Association. Clift suggested in a letter also sent in December 1871 that the federal government should oversee the production of American shad, but in this case for introduction into the waterways of the Mississippi River basin. Clift's idea dovetailed perfectly with the joint resolution: introducing shad into interior streams would create an additional stock of food fish for the westward expanding population. This project was also likely to get favorable attention from those politicians representing the several states involved, thus raising the likelihood of congressional approval. Unlike Edmunds, Clift even put a price tag of fifteen thousand dollars on his proposal.

Baird was a realist. Given his failure to provide a workable solution in the line versus trap fishermen dispute, how receptive would Congress be if he came asking for more money to experiment with fish propagation? What would taciturn congressmen like Farnsworth of Illinois and Benjamin of Missouri have to say? Baird gave Edmunds and Clift the same advice: lock down political support for their projects before trying to seek federal funding. Baird thought that the projects were interesting and potentially important steps in the process of rebuilding the nation's fish populations. But he also knew enough about the political machinery of Washington to know that for these projects to have a chance at being funded, they must be championed by others rather than just Baird. In both cases Baird suggested that if Congress was willing to fund either project, the Commission would gladly include it under its umbrella of activities.

And this is where the pivot occurred. The campaign by members of the American Fish Culturists' Association and like-minded individuals to involve the federal government in a program of fish propagation and stocking on a nationwide scale came to a crescendo in early 1872. As noted in the previous chapter, an important event in the history of fish culture in the United States occurred at the February 1872 meeting of the association (other than making

Baird an honorary member). George S. Page, a New York State fish commissioner, introduced a resolution to the association to petition Congress for enough money to fund both the Edmunds and Clift proposals. Page's resolution, you might recall, asked Congress to fund the building and management of two federally funded hatcheries, one near Puget Sound that would concentrate on Pacific salmon propagation, and another to be located on the East Coast that would propagate shad. Page's omnibus plan gave the Commission oversight of both.

Within a week of the association's meeting, Page was in Washington DC, visiting with Baird and fellow New York fish commissioner and first-term congressman Robert Roosevelt. Page's proposal and the involvement of Roosevelt and Baird to gain passage for the funding marks a critical point in the history of the Commission, and in the development of sport fishing in America.

Enter Robert Barnwell Roosevelt

Search "Robert Barnwell Roosevelt" online, and you're likely to come across a phrase that reads something like "originated the bill to create the United States Fish Commission." I cannot verify that statement, nor could a legal librarian at the U.S. Library of Congress. While he shouldn't get all of the credit, Roosevelt's actions in the spring of 1872 helped determine the future of the Commission.

The freshman congressman from New York had a busy day on 4 March 1872. He introduced one bill that dealt with smuggling (HR 1794), and another with government seizure of books and papers (HR 1793). Most important was his HR 1792, a bill "to authorize the erection of fish-breeding establishments." The *Congressional Globe* tells us that this bill was "read a first and second time, and, with the accompanying memorial, referred to the Committee on Appropriations, and ordered to be printed."[16] Roosevelt's bill thus introduced into the public arena and into the maws of the political machinery Page's dual-hatchery plan.

The bill requested that Congress appropriate ten thousand dollars (equivalent to more than a quarter-million today) to build and staff the two coastal hatcheries. Obviously this represented only seed money: surely the Commission would be back asking for additional funds to maintain and run these hatcheries into the future. (Remember Missouri Congressman Benjamin's warning?) The committee to which the bill was referred was chaired by Ohio Congressman James A. Garfield, who you might recall was implicated in the Credit Mobilier scandal and would become the short-termed twentieth president of the United States. (He would die from an assassin's bullet six months into his first term.)

Baird launched an intensive lobbying campaign to support Roosevelt's bill. He visited Congressman Garfield several times over the next few months, plying him with arguments about fish culture being the most workable answer to the coastal fish problem. Baird pointed out that Roosevelt's scheme could be used to promote the government's role in increasing the availability of food fish in all the states. Baird, always one to recognize the potential political interests at hand, pushed the fact that passage of the bill would result in shad being stocked in rivers in the Mississippi River basin, Garfield's political backyard. If the Commission's proposed activities created a shad run to the Gulf of Mexico (and back), everyone in that vast region of the country would benefit. After all, wasn't it the Commission's charge to investigate (read: solve) problems with the supply of food fish in interior states as well as coastal states? Getting interior-state politicians like Garfield on board would be critical for the passage of Roosevelt's bill, and for the future of the Commission.

Even with lobbying efforts of Baird, Page, and Roosevelt, the bill languished in committee. This may not be too surprising, given the initial reaction to the previous year's request for funds to support the Commission's coastal studies and the lack of any clear success in those efforts. To give his bill a boost, Roosevelt took to the floor of the House on 13 May 1872. There he delivered a speech that presented to his colleagues and to the public one of the most explicit and comprehensive analyses of how fish culture could solve the problem of reduced fish stocks—and, based on this reasoning, why wouldn't Congress vote for his bill?[17]

Roosevelt's wide-ranging speech included a history lesson in how and why fish losses in Europe and the United States arose. He delivered a fairly detailed history of fish culture and discussed the propagation techniques used by modern fish culturists. He even included an exposition on the probability of a naturally spawned fish versus an artificially propagated fish reaching adulthood. By the time he rounded the three-quarter pole, Roosevelt had reached the crux of his speech: if we depend on nature, his argument went, wild fish simply cannot reproduce fast enough to satisfy the growing human demand. If fish are tended to by humans (i.e., employing the new techniques of fish culture), we could produce an abundant supply of fish, enabling a few to feed the many.

Their reproductive power can only maintain a certain equilibrium; incline that toward destruction, and the entire class will quickly disappear. Treat them like wild animals, and they will inevitably be exterminated; domesticate them, as it were, encourage their growth by putting them under healthful influences, protect them from unseasonable disturbance, let them breed in peace, guard the young from injury, assist them by artificial aid, select the best varieties for appropriate waters, and we will soon augment the supply as greatly as we do with either land animals or vegetables.[18]

Roosevelt essentially presented the case made by the American Fish Culturists' Association: apply the techniques of modern fish culture and the stocks of wild fish could be restored. In addition, through human intervention the current range of fish species in the country could be expanded far beyond those currently imposed by nature. That is, if trout are not native to, say, Missouri or Iowa, simply use fish culture to introduce them. Reflecting the mindset of the day, human intervention based on modern science would save the day. But, he warned, the job would require the efforts of more than a few individuals or even states. This Herculean task required the federal government to intercede. Roosevelt opined that it "clearly [is] the duty of the [federal] Government to assist in this very work of introducing new varieties, as well as replenishing the old where they have been reduced.... This is the nation's duty or it is nobody's."[19]

Why should the solution lie in the hands of the federal government? The argument goes something like this: If we (fill in some state's name) impose restrictions that reduce overfishing, what good will it do if our neighbor (insert another state's name) does not? We will only harm our own citizens (i.e., the voting public), and there likely will be no improvement in the condition of the fish. This logic essentially exposes what occurred between the New England states in the line versus trapper fishermen dispute. Unless all states acted as one, no state would budge. In Roosevelt's plan the federal government would not trample on state's rights but "assist" states in this work: it would supply state commissions (or even private individuals) with fish and leave it to the local authorities to plant them where they saw fit. This, in fact, is how much of the Commission's eventual stocking program would be carried out.

If Roosevelt's argument swayed any of his fellow congressmen, he knew all too well that the sticking point would be the cost of this fanciful scheme. He tried to assuage such concerns by noting that "the cost of this undertaking is insignificantly moderate. A salmon-hatching house can be built for $1,000 while the necessary implements for shad raising are too inexpensive to be worth mentioning. Some labor must be employed, but it is mostly unskilled and cheap, while the outlay for transportation is simply the mere charge of express or traveling fare."[20] Besides, the net benefits to society, especially the less affluent, would be significant: "The people of this country would not grudge this were it a hundred times as great with the certain prospect of developing a new food resource and of diminishing the price of living to the poor."[21] To his fellow congressmen and especially anyone in the public who might read his speech, Roosevelt painted a rosy outcome for his plan:

> The mighty rivers of the southern and western States, which now produce generally only the poorer sorts [of fish], could readily be stocked with the most palatable and prolific sorts... Other rivers remain still unimproved, and several foreign species of fish should be introduced. For instance, the magnificent Danube salmon, which attains a weight of a hundred pounds, might be acclimatized in the Ohio and the upper Mississippi, while the true salmon might be brought to the Delaware and Susquehanna.[22]

Roosevelt argued and truly believed that the benefits of the proposed endeavor would improve the welfare of many across the country. Not only did recent findings of modern science support the undertaking, but the transportation infrastructure needed to successfully achieve such lofty goals already existed: the railroad system. Indeed, Roosevelt noted that trout spawn (and presumably any other species) could be "sent from one end of our country to the other with as little trouble or danger as letters... delivered by express [trains] precisely as any other packages."[23]

Roosevelt seemed to be daring his colleagues to not pass his bill. What politician would want to be shamed as begrudging his constituents of this piscatorial windfall? And at no cost to the state? The public record has little to say about Roosevelt's speech. Those in the fish culture fraternity understood its importance and expressed their gratitude. William Clift's presidential address at the 1873 American Fish Culturists' Association meeting convened in New York City praised Roosevelt's public mindedness, proposing that he "deserves the thanks of this association *and of all patriotic men* for these appreciative words spoken in his place in the House of Representatives, in behalf of this movement for stocking the barren rivers." (emphasis added)[24]

Though historically important, Roosevelt's speech did not sway his congressional colleagues. His bill never even made it out of committee. And while Garfield tried to back-door passage Roosevelt's proposal by placing it into a Sundry Civil Appropriations bill before the House, some felt that the proposal for the fish hatcheries represented new legislation and should not be included in the appropriations bill according to House rules.[25] The fight was not over, however. Baird persuaded Senator Edmunds of Vermont to insert language into an amended version of the Sundry Appropriations bill before the Senate that requested the funds—even increasing them to fifteen thousand dollars from the original ten thousand—for the introduction of shad, salmon, whitefish, "and other useful fishes in the waters of the United States to which they are best adapted."[26] An edited version omitted any reference to building hatcheries and increased the geographical scope of the proposed activity.

These changes worked. On 1 June 1872, the Senate Appropriations Committee sent the amended appropriations bill to the floor, where it passed without dissent. Senator H. W. Corbett of Oregon slightly modified the proposal to include the introduction of shad to waters along the Pacific Coast. A week later, with that minor change, the bill went back to the House, where the Sundry Civil Appropriations Act was passed. In essence the House passed the proposal they had earlier rejected and gave the Commission even more money to carry out what they wanted to do.

Passage of the Sundry Appropriations Act in the summer of 1872 gave Baird and his Commission the money to fund the construction of the two fish hatcheries. Indeed, these would be the first of the Commission's many federal fish hatcheries spread around the country. More importantly, the Commission now had the green light to embark on the most comprehensive and far-reaching experiment in fish culture ever attempted in this country.

A Final Note

It is important to keep in mind that fish culturists and their supporters believed that humans held the power to bend nature to their own will and desires. They could, with the application of science, manage fish populations in any body of water as they saw fit. This mindset is illustrated by the observation made by William H. Ludlow, chief engineer to Lieutenant Colonel George A. Custer's July 1874 expedition of the Black Hills. Upon finding the headwaters of Castle Creek Ludlow remarked that "we were continually looking for trout in these streams, which seemed as though made expressly for that fish." Not finding any, Ludlow offered a simple solution: "There could be no finer trout-streams in the world than these were they once stocked."[27] If western streams were "barren," as they often were described, why not simply stock whatever species (preferably trout) you thought should be there?

And so, in the early 1870s, the new U.S. Fish Commission initiated its great experiment to test the hypothesis that propagating and transplanting fish across the country would solve the decline in fish stocks. As one historian would suggest, the Commission represented "the first federal agency to deal with the conservation of a specific natural resource."[28] Combining forces with the railroads that were crisscrossing the country, the two would forever change the nature of the country's fishery and of sport fishing.

4

THE FISH COMMISSION, THE GREAT EXPERIMENT, AND THE RAILROADS

The Commission was tasked with finding some way to repopulate the nation's streams and lakes. If successful their actions would increase food sources for the growing population. This goal would be an overriding theme in the Commission's work through the rest of the 1800s. Vermont fish commissioner M. C. Edmunds even called upon his fellow members of the American Fish Culturists' Association to join, what he considered, a crusade. "Are we not," he asked, "commanded 'to feed the hungry,' and how better can this great duty be performed than by laboring to restock our lakes and rivers with fish of all kinds?"[1] How in the world would the Commission accomplish this monumental task?

Origins of the Great Experiment
What I am calling the "Great Experiment" amounted to figuring out which fish might best grow where. For example, if American shad from the East Coast were introduced into Western and Midwestern waters, would they survive and increase in numbers to such an extent that they would become a reliable food source? No one knew the answers to such questions, but they were all hypotheses that the Commission would test.

Once the Sundry Appropriations Act passed, Baird quickly arranged a meeting in Boston in June 1872 with representatives of several eastern state fish commissioners and members of the American Fish Culturists' Association. The top agenda item was how best to distribute the Commission's recently allocated funds (fifteen thousand dollars) to best satisfy the parameters of the joint resolution. The group agreed that it would be best to split the funds in three ways. One-third would go to building and maintaining a salmon hatchery on the Penobscot River in Maine. Charles Atkins would head up this project. Given the real and political resources spent thus far in dealing with the coastal fisheries problem, it remained important—scientifically and politically—to see if Atlantic salmon could be successfully propagated and returned to eastern waters. Though important for the Commission's objectives, this aspect of the experiment has less importance for the rest of this story, so I will not discuss it any further.[2]

Reverend William Clift, the first president of the American Fish Culturists' Association and a Connecticut fish commissioner, sent along a letter to be read at the meeting. He argued that a portion of the funding should go to stocking the Mississippi River and its tributaries with shad. The Commission also should try and establish eastern shad in West Coast rivers. Those at the meeting thought these projects would be politically popular, because many in Congress who voted for the allocation came from states in the Mississippi River Valley. This idea passed, so another third of the funding was earmarked for propagating and distributing eastern shad. Clift and New York fish culturist Seth Green were chosen to oversee the shad initiative.

The third project was, I think, the most audacious. The question to be answered was: "Could Pacific salmon be introduced into eastern waters to augment or replace the depleted stocks of Atlantic salmon?" To answer this question the Commission would need to build a hatchery in Northern California or Oregon—they were not sure exactly where. At this hatchery salmon eggs would be collected, fertilized, and sent to Commission hatcheries in the East, where the eggs would be "hatched," and the resulting fry deposited into streams in states along the Eastern Seaboard north of Washington DC. Basically, it was the shad project in reverse: eastern shad were going west while Pacific salmon were going east. Another similarity with the shad program entailed the planned release of Pacific salmon into rivers throughout the Mississippi River system. This allowed the Commission to test if the salmon would create a spawning run from states like Ohio or Missouri to the Gulf of Mexico and back. One-third of the Commission's funding was allocated for this bold— some might think foolhardy—experiment. The thirty-five-year-old Livingston Stone was tapped to supervise the salmon experiment.

With the funds allocated and the responsibilities assigned, work soon began on each. In what follows I will provide an overview of the shad and salmon operations. Because these activities soon morphed into the transplanting of other species, most notably rainbow trout and carp, I will cover those aspects of the Commission's experiment as well.

Shad

American shad—more specifically, *Alosa sapidissima*—had been a dietary staple for folks living on the East Coast for decades. And, even though the line versus trap fishermen debate may have given the impression that it was just a northeastern fish, shad are found along the entire Eastern Seaboard, from the Saint John's River in Florida north to the Saint Lawrence River. It represented an important livelihood for many who were engaged in the business of catching and selling them. Though the shad's spring-summer spawning runs had declined significantly over time, could the science of fish culture rectify that problem?

Baird recognized in his *Report of the Commissioner for 1872 and 1873* that 1867 marked an important year in the effort to manage the shad population: Seth Green, working for the state of New York, had introduced several innovations that made stocking shad on the enormous scale envisioned by Baird possible. Green developed a hatching box that you would float in the river from which the shad eggs had been taken. The design of the box allowed the eggs to be in the same water as if they had been laid by females. The box even created a slight eddy that gently stirred the eggs. Green's innovation proved so successful that millions of shad soon were being hatched and stocked by several state fish commissions in the Northeast. His success also raised the interest of other states. When the transcontinental railroad opened for business, the California State Fish Commission ordered a shipment of shad fry.[3] Green accompanied this shipment from New York to California and, along with members of the California commission, deposited them into the Sacramento River. The first bicoastal transplant of fish had become history.

If states were already engaged in trying to rebuild the shad runs, why would the Commission get involved? One reason is that the federal government could bring to bear more resources than any one state. In addition, the Commission looked beyond the state level. It was unlikely that, say, Ohio alone would bear the cost and risk of stocking its streams with shad to create a spawning run to the Gulf if individuals in every other state between Ohio and the Gulf could harvest the shad on their return trip. Besides, these boundary waters were deemed property of the federal government, so success would benefit a broader population (and the Commission) than any one state.

The Commission thought the rivers of the Mississippi River Valley were appropriate laboratories for their experiment. The scientific side would answer the question of whether shad could generate new spawning runs through the Mississippi River system to the Gulf of Mexico. The practical side might just make good on Edmund's admonition to "feed the hungry" and the joint resolution's charge to increase the nation's stock of fish. Though referring to the salmon experiment, Livingston Stone captured the general sentiment of the undertaking when he recognized that the Commission's efforts might "undoubtedly, be a total failure, but should the Commission make a success of a single river of a size . . . it would pay for all that has been expended in the direction of all the waters of the United States."[4] Think of it like playing most lotteries: the probability of winning is very low, but if you do not play your chance to win the jackpot is zero. So why not at least try?

The "success" of the Commission's experiment also had as much to do with satisfying public and political expectations as with actually repopulating lakes and streams. Baird was blunt about this: "The object is to introduce [fish] into as many waters as possible and have credit with Congress accordingly. If they are there, they are there, and we can so swear, and that is the end of it."[5] Did depositing fish in as many political jurisdictions as possible across the country outweigh the actual success producing a viable fish population? Baird, ever the political realist, recognized that grateful congressmen would be more inclined to vote for his annual budget requests, the bulk of which went to fund the Commission's fish culture work.

In 1872, the Commission's first year, the shad operation produced only about 7.5 million fry. That may not seem small but wait until you see the numbers in future years. Green focused on stocking rivers in the Northeast, the majority of shad being deposited into the Hudson River. Other sites included the Genesee River; Oneida Lake, which connects to Lake Ontario; and Lake Champlain, which empties into the Saint Lawrence River. Keeping with the plan hatched at the Boston meeting, Green sent some of his employees to St. Paul, Minnesota, where they deposited twenty-five thousand shad into the upper section of the Mississippi River.

The shad fry often were transported in cans, standard dairy milk cans loaded in railroad baggage cars. This, as you will see, was the standard technology of transporting fry, one used throughout the second half of the 1800s. Accompanying the cans of fish were one or several Commission employees—known as "fish messengers"—assigned to oversee the fry en route and to deposit them in their new homes.

This process of shipping fry might sound simple enough, but it actually was quite challenging. Let me use an early trip by William Clift in 1872 to illustrate.[6] Clift started off with ten eight-gallon cans containing a total of about two million shad fry (shad fry are very small). In the early morning hours of 2 July, Clift and his assistants loaded their cans of fish onto the Erie Railroad at Holyoke and headed for Salamanca, a small town in western New York. They reached Salamanca about twelve hours later. Clift's orders from Baird were to deposit some (exact numbers were not needed) shad into the Alleghany River. Salamanca represented an ideal point to do this because the railroad station was close to the river. The fact that railroads often followed rivers or crossed them will be a recurring theme throughout this story: many times, fish would be deposited into this or that river or stream simply because the railroad crossed it or ran near it.

Clift carried out his instructions at Salamanca and deposited about four hundred thousand shad fry into the Alleghany River. Actually, he "left here three cans ... in charge of the ticket-master, who promised to see them immediately put into the river." This will be another recurring theme: railroad employees—in this case, the ticket-master for the Erie Railroad—often were called upon to carry out the Commission's work of stocking fish. And it wasn't necessarily because the employees were altruistic. As you will see later, some railroads made it company policy.

At Salamanca the remaining seven cans of fry were transferred to a baggage car of the Atlantic & Great Western Railroad bound for Indianapolis, where Clift planned to make a large deposit of shad into the White River. At a stop in Kent, Ohio, with the railroad station again conveniently close to the Cuyahoga River, Clift deposited some shad into that river. (As an aside: this rather haphazard approach to where shad, or any of the species discussed later, were introduced is a hallmark of the Commission's approach. It also explains why the Commission's work had longer-term, negative repercussions on native species.) Reaching Indianapolis on the nation's birthday, Clift missed the connecting train to St. Louis. The delay—some ten hours—meant that more of the shad could expire if penned up in the cans.

At Indianapolis Clift decided to take a gamble. Near the railroad bridge spanning the White River just outside of town, he planted several cans of shad fry. But he also kept several other cans to test how far fry could be shipped without significant loss. "We thought," Clift wrote, "that if we could carry them [the remaining shad fry] across the plains and plant them in the Platte [River] at Denver it would be demonstrated that all the streams in the country can be furnished with shad fry." After all, Clift argued, "the experience would be worth all it cost, even if we failed." So, Clift and his charge of shad left Indianapolis for Denver, with planned stops in St. Louis and Kansas City. At stops in Washington and Herman, Missouri, sixty and eighty miles, respectively, west of St. Louis, Clift stocked the Missouri River with some of the remaining shad. When he reached Kansas City the last can, which contained about two thousand shad fry, was loaded on to the Kansas Pacific Railroad bound for Denver. When Clift finally reached Denver on the morning of 7 July, he found that the remaining shad had survived the trip in relatively good condition. These long-distance travelers were released into the Platte River, far from their birthplace. This, reported Clift, gave him the "joy of seeing Connecticut River shad swimming in the waters of the Platte."

Baird understood that the success of the shad program would depend on the public's recognition and approval of the Commission's efforts. Green and Clift's stocking trips—and they were not exceptional—relied on the railroads. Baird thus published a public "thank you" to the railroads in his *Report* for 1872–73. In his words, "Very valuable assistance was rendered ... by the express companies, especially the Adams and the American and Merchants' Union. Without the help of special instructions to their agents to assist Messrs. Green and Clift, it would have been difficult to accomplish the object in view."[7] This public recognition of the railroad's assistance would be another constant in the Commission's ongoing experiments. It would also occur at the state level, once state fish commissions started conducting their own fish transplanting experiments.

Clift's descriptions make this journey sound quite simple, but transporting shad fry (or any other fish, for that matter) in milk cans was anything but. Shad are especially finicky. They required the water to be constantly checked for clarity and a narrow temperature range maintained. Aside from the shad's predilections, a milk can loaded with fry and water is, to say the least, burdensome. A loaded can tips the scales somewhere between seventy and ninety pounds, depending on how many gallons of water it holds. And if you recall the description of long-distance railroad travel requiring the transfer of packages between the many train lines that a trip of any distance likely required, think of hauling a bunch of water-laden milk cans from one train to another through busy railroad stations.

After a couple of years into the shad experiment, Baird was so encouraged by its success—in terms of being able to transport the shad fry long distances by rail—that he saw "no reason why any stream in the United States having direct communication with the Gulf of Mexico, or either ocean, may not be made to abound in an equal degree with these [shad] and other fishes, and in view of the aggregate of the animal food to be derived from a number of such streams, the importance of this work can hardly be overestimated." Baird went on to argue that the Commission was fulfilling the objective of the resolution. "The operations of the Commission," he averred, "have ... given entire satisfaction to the people at large, as shown by the general popularity of the measures adopted, the great interest excited in the subject throughout the country."[8] And if the public were pleased, so, too, would be the local politicians and those in Washington DC who vote to fund the Commission.

Once the operation was up and running, hundreds of thousands of shad were being shipped all across the country under the watchful eye of Commission employees. Reports of shad deposits like Clift's can be found in the *Report of the Commissioner* for nearly every year through the end of the century. The Commission planted shad in lakes and streams, the Great Lakes and their tributaries, and even in the Great Salt Lake in Utah. If shad could survive in a variety of environments, the Commission's plantings would reveal where. And the more places the better: shad could become an important source of dietary protein for more and more people.

Instead of repeating the logs of the dozens of trips made to deposit shad over the remaining years, let me use a few numbers to illustrate the magnitude of the shad experiment. Over the 1870s the number of shad deposited (nationally) fluctuated between a low of three million in 1874 to a high of twenty million in 1880. The variance reflects the fact that producing shad fry for distribution was still influenced by the vagaries of nature. In total, between 1872 and 1880 the Commission stocked over eighty-four million shad into lakes, rivers, and streams across the nation. And the program only grew after that. In 1885 alone the Commission shipped nearly thirty-five million shad fry around the country. By 1900 the *annual* shipment had grown to over two hundred million fish. Keep in mind that these numbers are only for the Commission: a number of state fish commissions also were engaged in their own plantings, so the true number of shad being shipped and deposited actually is much larger. And, remember, shad was only one of the species that constituted the Commission's Great Experiment.

The geographical scope of the shad program grew with the number of shad transported. Shad were being deposited into nearly every state by 1880. Much of this experimentation pushed the boundaries of fish culture, such as placing shad into the Great Salt Lake. But Baird always promoted the practical side of the Commission's work, even where it may have seemed foolhardy. The commissioner pointed to the guidelines of the joint resolution as his guide: replenish the depleted shad population in those states where they are native, and introduce them in states where they might thrive. Introducing shad into, say, Indiana or Kansas, or even the Great Salt Lake, fit the objective of trying to increase the supply of food fish for as many in the population as practicable. In the end the shad experiment can be considered a success story: today the Columbia River holds one of the largest populations of American shad.

In all of this, railroads were the necessary condition for success. They made it possible for the fish messengers and their cans of fish to travel about the country on a web of track that gave the Commission access to a wide range of rivers and streams. The railroads bought into the Commission's activities to such an extent that it was not uncommon for an engineer to stop the train so a messenger—or a railroad employee—could dump a can or two of fish into the stream nearby. This relationship was not unique to planting shad: as you will see, it extended to the Commission's other experiments as well.

Salmon

Recall that attendees of Baird's meeting with the American Fish Culturists' Association in Boston agreed to a plan to transplant Pacific salmon into eastern rivers. Even though there was broad agreement to the salmon experiment, uncertainties clouded its success. For example, those in attendance—all fish culturists of some repute and the U.S. fish commissioner—were not exactly sure which species of Pacific salmon would be best suited for this undertaking. Another question that arose concerned the aptitude of Pacific salmon to, as fly fishermen would say, "take the fly" like its Atlantic cousin. The catchability of the Pacific salmon was important, because it raised the question of whether Pacific salmon would be welcomed in the East as a food fish and a sporting fish. Another unknown: Would Pacific salmon breed with the Atlantic salmon? (They won't.) And no one seemed aware of or too concerned about the fact that Pacific salmon die after spawning (they are *semelparous*) unlike the Atlantic salmon which returns to the sea to repeat the spawning cycle (they are *itero-*

parus). Even with such uncertainties—some, it seems to me, quite relevant for the success of the experiment—the Commission pressed on.

Soon after the Boston meeting, Baird wrote to Livingston Stone in July 1872 to officially bestow on him the title of deputy fish commissioner. That and to inform Stone that he should proceed "at the earliest possible moment" to California to begin the process of collecting salmon eggs. Baird suggested that as soon as Stone arrived in San Francisco he should "by examination and counsel with those who are familiar with the subject" determine the appropriate species of salmon best suited for the Commission's purpose. Baird also offered no counsel as to where Stone should set up his salmon-taking operation. One thing that Baird was clear about, however, was Stone's mission: "Lay the foundation of an arrangement, on a large scale, for obtaining eggs of the best varieties of *Salmonida*; and other food-fishes of the western coast."[9]

Stone wholeheartedly accepted the position. And Baird could not have asked for a more enthusiastic adherent to the plan. As questionable as the venture seemed from the outset, Stone expressed no qualms about its potential. "With the vast resources of Pacific rivers, salmon eggs should be made almost as plenty as water," Stone wrote in his reply to Baird. He believed that the experiment "will best satisfy the people for their expenditure. It will also relieve in a great measure the oppressive restrictions, which must otherwise be placed on the fishing of the restocked rivers ... the enterprise ... I believe to be the one great work above all others in the restoration of salmon to the American Rivers and lakes."[10] His misgivings would come later.

Stone wasted no time in undertaking his new responsibilities. He put his beloved Cold Springs hatchery up for sale and departed for Boston on the first of August. In Boston he boarded the first of many trains to make the still novel cross-country trek to San Francisco. Think of it: only a few years earlier there simply was no train to San Francisco, and now Stone set out from Boston for California to undertake one of the grandest experiments in the history of fish culture.[11]

As soon as he arrived in San Francisco, Stone sought advice on the spawning habits of Pacific salmon. His immediate goal was to establish an egg collecting station up the Sacramento River—a major arterial for spawning salmon. If that should prove unsuccessful, Baird suggested that Stone explore potential sites as far north as the Columbia River in Oregon. Wherever he chose to open shop, Stone needed to act quickly: the salmon run would soon be underway, and failing to quickly find a suitable location would delay his operation by a year, a possibility that he (and Baird) considered unacceptable.

Stone met with several individuals without discovering where to find spawning salmon. Luckily, a California Commissioners of Fisheries board member introduced Stone to the chief engineer of the Southern Pacific Railroad, a Mr. Montague. Montague shared with Stone the company's survey maps of the upper stretches of the Sacramento River. On one of these Montague pointed out a stretch of the river near the junction of the Pitt and McCloud Rivers where he had seen salmon during previous spawning runs. Based on Montague's recommendation, Stone soon boarded the Southern Pacific bound for the line's northernmost station at Red Bluff, California. From there he continued northward on horseback and soon found *the* spot. Here on the banks of the McCloud River, the local Native Americans—members of the Wintu tribe—had been harvesting spawning salmon for many years. Stone decided there and then to build the government's first salmon hatchery on the West Coast. Like many other aspects of Western development, members of the local Wintu tribe were not consulted.[12]

The site on the McCloud River was ideal for the purpose at hand. Because it flowed from the base of Mount Shasta, underground springs ensured a uniform flow of cold, clear water. Another reason for going so far north is because by this time many of the Sacramento's downstream tributaries already had lost their salmon runs.[13] Similar to the reasons affecting spawning runs in eastern rivers, the causes out west stemmed mostly from industrialization. Logging companies were clear-cutting the hillsides, causing erosion and the silting of the rivers. Mining companies using hydraulic mining techniques also fouled the rivers. And the blasting and clearing for railroad tracks and bridges generated runoff that polluted the rivers and streams. All these contributed to the end of the salmon runs.

Another reason to focus on the salmon run farther up the Sacramento River is because during the spawning season the river downstream often ran turgid and warmish. Such conditions enhanced the possibility that Pacific salmon might be adaptable to eastern rivers that had similar conditions. Moreover, if salmon fry released into streams fed by cold-water springs in interior states like Missouri that eventually drained into the Mississippi River, they would face similar conditions as they had for a millennia migrating up and down the Sacramento River. That is, would they migrate down the Mississippi River to the Gulf of Mexico and return to spawn in those interior streams where the initial deposits were made? And, even though it just sounds like a nonstarter, the Commission had conducted tests to determine if the water temperatures at low enough depths of the Gulf were cool enough for salmon to survive. With evidence suggesting that it was, salmon spending a part of their lives in the Gulf before returning to Missouri did not seem too far-fetched. So, like the shad experiment, why not test the hypothesis? If salmon could be acquired at low enough cost—Stone's assignment—it seemed foolish not to try. If the outcome could be restoration of the Atlantic fishery and the creation of new one in the central part of the country, who would say no?

Stone, along with two assistants who followed him to California—Myron Green and Stone's nephew William T. Perrin—built their rudimentary salmon-taking station in early September. With a simple hatchery built on the banks of the McCloud, they immediately set about to capture spawning salmon, collecting eggs from the females and milt from the males. The process of harvesting the eggs was the same as Coste had described years earlier: hold the females with their tails toward a pail on the ground and squeeze out the eggs using a pressing, downward motion along their abdomens. Milt from the males was obtained in a similar fashion. The eggs and the milt were mixed together with some water, and after a short period of time the fertilization process was complete. The now-fertilized eggs were then placed into trays to await their development. After the fertilized eggs reached a certain degree of ripeness, the plan was to pack them up and ship them to hatcheries back east, where the fry from these eggs would be transplanted into nearby streams and rivers or shipped to others for deposit.

Due to their late start in the spawning run, Stone's modest hatchery was matched by a modest production of shippable eggs. On 25 October 1872, just about two months after Stone had arrived in California, a batch of thirty thousand Pacific salmon eggs (more specifically, Chinook salmon) were sent eastward.

How were the eggs shipped? The fertilized eggs—each about one-quarter of an inch in diameter—were packed into wooden boxes that measured two feet square by one foot deep. Each box was filled in the following manner: first a layer of moss was seated with mosquito bar (netting) on top of that. A layer of eggs was spread across the netting, followed by another layer of netting, moss, more netting, another layer of eggs, and so on, until the box was half-full. At this point a horizontal wooden partition was put in place to reduce the likelihood that the upper layers of moss and eggs would squash those toward the bottom. With the partition in place, the layering process began anew until the box was filled and a lid screwed to the top.

Two of these egg boxes were fit into a larger crate, making sure to leave a space around the outside of the boxes. Into this space moss and ice were placed with the hope that the combination would keep the eggs at the desired temperature, regardless of the temperature outside. A final layer of ice was placed on top of the boxes before screwing on the lid to the crate. With this manner of packing the eggs, each crate could hold upward of several hundred thousand eggs and weigh about three hundred pounds.[14]

In late October the crate of salmon eggs left the McCloud hatchery on a horse-drawn wagon over rough mountain paths and dirt roads to a rail site twenty-five miles away. Here the crate was transferred to agents of Wells Fargo & Company and booked for a through railroad trip to their final destination in the East. The first leg of the train ride was to Sacramento, where the crate was loaded onto a freight car of the Central Pacific attached to a Union Pacific train bound for Omaha. At Omaha the crate again would be transferred to a train heading to Chicago, where it would be transferred several more times between different train lines before reaching its intended destination, Bloomsbury, New Jersey.

Think of it: each crate weighed several hundred pounds, making them, to say the least, difficult to handle. Unlike the cans of shad fry hauled by train, no fish messenger supervised the crate's journey. Instead railroad employees attended to all of this transferring between railroads, making sure that the crate arrived on schedule. When the crate reached Bloomsbury, workers at the hatchery discovered that only six thousand out of the original thirty thousand salmon eggs were viable. From these only a couple of hundred fry were successfully hatched. Not a glorious beginning, but not a total failure either. Stone had at least demonstrated that fish eggs could be transported thousands of miles. Of course, without the railroads, even getting this far was unimaginable.

The original plan involved stocking these surviving Pacific salmon fry into Pennsylvania's Susquehanna River. An unusually cold fall followed by especially nasty winter weather delayed the initial deposit. It wasn't until March 1873 that these young, pioneering Pacific salmon were released into the Susquehanna near Harrisburg. Though the in-transit survival rate was low, and the final deposit took some time, the Commission could rightfully claim that it could harvest eggs from Pacific salmon and deliver them to the East, where the fry would be planted. Perhaps the experiment to build up the salmon population might work after all. Of course, that was assuming the fry would survive in their new surroundings.

The number of eggs shipped increased sharply in the next year, to about 1,400,000. Shipping this many eggs required five different allotments. The first left the McCloud station on 20 September, and the last departed on 10 October. To get a better idea of the difficulties involved, let's use one trip, when Stone accompanied a shipment to Boston of three crates, a total of about 500,000 eggs. The trip is illustrative of the difficulties involved with transporting these large and heavy crates. (Note especially the time stamps.) What follows is an edited version of Stone's travel log.[15]

> 30 September 1873 (Tuesday)
> > Depart McCloud hatchery at 4 p.m. and head to railroad station at Redding, CA, by wagon; arrive 1 a.m.
>
> 1 October (Wednesday)
> > Train leaves Redding 3:00 a.m. for Sacramento; arrives 1 p.m.
> > Crates loaded into Wells & Fargo express-car, on Central Pacific Railroad.
> > Leaves station at 2 p.m.
>
> 2 October (Thursday)
> > Train reaches Carlin, Nevada in morning. Crates re-iced.
>
> 3 October (Friday)
> > Reach Ogden, Utah at 7 a.m.
> > Crates transferred from Central Pacific Railroad to express car of the Union Pacific Railroad.
> > Crates re-iced; eggs checked and in perfect condition.
>
> 4 October (Saturday)
> > It was necessary to constantly check the egg crates to make sure the temperature was kept fairly constant. As Stone relates, "Went to express-car to examine the eggs" and found that the "express-manager [railroad employee] had kept a hot coal-fire in the car and it was very hot." This meant an unexpected need for more ice, which was in short supply.
> > Ice procured at Cheyenne, Wyoming and crates re-iced; telegraphed to Laramie to have more ice available at that stop.
>
> 5 October (Sunday)

Reached Omaha, Nebraska at 1 p.m.

Crossed Missouri River reaching Council Bluffs, Iowa where express car transferred to the Chicago, Burlington and Quincy Railroad.

Left at 3 p.m. for Chicago.

6 October (Monday)

Reached Chicago at 3 p.m.

Union Pacific express car transferred to the Michigan Central Railroad.

Left Chicago at 5:15 p.m.

7 October (Tuesday)

Entered Canada on the Great Western Railroad at 4 a.m.

According to Stone, at this point "the Union Pacific express-car, which still accompanied the train, was sealed up by the custom-house officers, so that I could not enter it till we left Suspension Bridge that afternoon at 2 o'clock." It so happens that the car was carrying a load of gold and silver bullion. Luckily, or perhaps with forethought, Stone had with him a letter from one of the head managers of Wells & Fargo's express and with it gained access to the express car.

Arrived Rochester, New York at 5 p.m.

Left three boxes (200,000 eggs) for Seth Green.

Left Rochester and arrived at Albany, New York at 2 a.m.

8 October (Wednesday)

Train arrived in Boston, Massachusetts at 8 a.m.

From the McCloud River to Boston, the egg crates covered nearly four thousand miles by rail, required the services of five different railroad companies, and lasted nine days.[16] Though quite a technical feat, the trip did not end well for many of the salmon fry-to-be. When Stone went to bed in Albany, he thought that the express car carrying the crates would be attached to the Boston-bound train. Unbeknownst to Stone, while he continued to Boston the express car remained at Albany. "To my great surprise and dismay," an abashed Stone wrote, "I could not find the salmon eggs for Mr. Atkins and Mr. Brackett, and now learned for the first time that they had been left with the car in Albany. I was the more chagrined at this because I had been so very careful to keep with them." It was like the modern-day version of lost luggage on a multistop airline trip. Stone "could not get track of these eggs again or learn for some time what delayed them; and it was three days before Mr. Brackett got his and four days for Mr. Atkins received his."[17] Needless to say, the egg loss for the Brackett and Atkins shipments was quite high.

Difficulties aside, the fact was that salmon eggs could be harvested, fertilized, and shipped in massive quantities to hatcheries back east. Stone wasted no time ramping up production at the McCloud hatchery. Soon the annual harvest of salmon eggs reached nine million. This required shipping seventy-nine crates weighing a total of nearly ten tons. Stone suggested that when packing the crates "they should be compactly arranged, in order to reduce the express charges, which are enormous at best."[18] Even though Wells & Fargo charged the Commission a below-market rate, it amounted to about three thousand dollars, nearly eighty thousand dollars in modern terms. Though Wells & Fargo charged the government to ship the crates, the records indicate that personnel attending the crates rode for free, either in the baggage car or with the other passengers.

The volume of salmon eggs taken and shipped from the McCloud station—Stone renamed it the Baird Hatchery in 1878—increased over time with improvements to the hatchery and with innovations in how to catch the salmon. Salmon egg production and shipment reached its peak in 1878, when fourteen million eggs were shipped. From that point on annual salmon egg production at the hatchery began to decline for various reasons. In one year massive rains flooded the hatchery, destroying some of the equipment. The Central Pacific's extension of its line north of Redding also reduced egg production. Because the extension followed the Little Sacramento River, blasting the hillsides to prepare the rail bed washed soil and debris into the river. This killed many fish, including migrating salmon. In addition the abundance and ease of harvesting the spawning salmon provided a readily available and free source of protein for the railroad's construction crews.

Stone's salmon-taking operations were temporarily suspended in 1884. This reflects the fact that the salmon experiment was failing. I will be more specific about this later. Ironically, production would resume with a different objective: propagating salmon for restocking *West* Coast rivers.

The story thus far is impressive simply based on the sheer numbers of eggs procured and shipped by the railroads to awaiting hatcheries, both those managed by the Commission and state-owned hatcheries. But the railroads' role in the experiment did not end there. Once the eggs reached their designated hatcheries and were hatched, the resulting fry were distributed using the same process as shad fry: loaded into milk cans and moved about in baggage cars with attending fish messengers. Once the shipments reached the designated train station, the fry were transferred to whatever contrivance the individual or organization who ordered them brought. If the recipients did not bring their own containers, they were allowed to use the Commission's cans, the proviso being that they would be returned to the station, where the stationmaster would see that they were sent back to the Commission, the understood agreement being that the railroad employee would take care of the details, gratis.

I noted earlier that during the 1870s the Commission deposited more than eighty-four million shad across the United States. How did the salmon stocking compare? The popularity of the project is evidenced by the fact that between 1872 and 1880 the Commission planted over twenty million salmon fry—all of which originated from eggs shipped from the McCloud—into the waters of nearly every state, even in seemingly unlikely states like Nevada and Mississippi. The top five states, in terms of numbers of fish deposited, were Maryland, Pennsylvania, New Jersey, Michigan, and Connecticut. These states accounted for about half of the total number of salmon planted. For this group, except for Michigan, the salmon fry were stocked into rivers that eventually flowed into the Atlantic Ocean. This pattern is consistent with the goal of trying to bolster or supplant the Atlantic salmon population. In Michigan it represented an attempt to create a spawning run to and from the tributaries of Lake Michigan and Lake Superior. In the end neither would be successful.[19]

One aspect of the salmon operation, unlike shad, is the fact that all of the salmon eggs originated from one source: the McCloud River.[20] How could so many states release salmon when Stone was only shipping impregnated eggs? During this time the Commission began building a system of federal fish hatcheries across the country. Salmon eggs shipped to those dispersed hatcheries were hatched and the salmon fry distributed to applicants, whether they be individuals or state fish commissions. It was also true that a number of states were establishing their own cold-water hatcheries. These salmon eggs were sent and hatched with the resulting fry stocked by the state commission in their rivers and streams. The whole process, whether federal or state, always relied on the railroads to move the goods.

It soon became apparent that adult Pacific salmon were not returning to spawn in the streams into which they were planted. More than likely, few if any survived for very long after being stocked. Whatever the reason the salmon experiment looked like a complete failure. Stone admitted that "except in very rare instances, these millions of young salmon were never seen again." Stone eulogized the millions of salmon he helped send to their demise with an air of melancholy: "What became of them? Where did they go? Are any of them still alive anywhere in the boundless ocean? Or are they all dead? And if they are dead, what killed them?"[21]

A decade after it began, the days of shipping viable salmon eggs from the McCloud station to hatcheries in the United States and to recipients around the world came to an end. Salmon egg shipments to hatcheries back east stopped in 1883. As I noted, the McCloud hatchery continued to propagate salmon for deposit in rivers along the Pacific Coast. In 1900, for example, the Commission shipped over 19 million salmon eggs and fry (mostly fry) to recipients in the states of California, Oregon, and Washington out of a total of nearly 20 million eggs collected. For some unexplained reason, 1,350 also were sent to the federal fish hatchery in Neosho, Missouri. Ironically, the program of producing salmon for introduction into eastern rivers had turned to one of producing salmon for reintroduction into the Pacific Northwest to offset the losses due to overfishing and the encroachment of civilization.[22]

The Commission's interest in salmon production to stem losses in the Northwest expanded in the mid-1880s. In 1886 Oregon's two senators, John Mitchell and Joseph Dolph, approached Baird to see if the Commission would assume management of the state's hatchery on the Clackamas River. Baird agreed with the condition that the Commission would receive free transportation for personnel and material on the two railroads that serviced the Northwest: the Northern Pacific Railroad and the Oregon Railway. The two senators negotiated on Baird's behalf with the railroads, which agreed to his conditions.[23] The Commission then took control of the hatcheries and continued to produce salmon for deposit in rivers of the Northwest.

The salmon experiment had, by all accounts, failed miserably. Even so it took until the late 1880s before the plug was officially pulled. Marshall McDonald, who had become fish commissioner in 1888 following Baird's death in 1887, drove the last nail into the coffin, writing that "in no single case did the experiment prove satisfactory." McDonald saw no other alternative but to "abandon an experiment which, reasoning from a priori considerations, gave fair promises of success."[24] With the salmon experiment a failure, the Commission needed to restore its reputation in the eyes of the public and the politicians who funded the Commission's activities. Luckily for the Commission, two fish would resuscitate that reputation. One of those fish is the rainbow trout.

Rainbow Trout

I mentioned earlier that the Great Experiment started with shad and salmon and would morph over time to include other species. It had become abundantly clear to the Commission and to the public that the salmon experiment was a bust. So, as early as 1879, the Commission pivoted once more. Baird instructed Stone to find a suitable location on the McCloud River where he could collect the eggs of another Northern California native from which, like the salmon, its eggs would be collected, fertilized, and shipped east. Some called it the "black spotted trout," and others called it the "California brook trout." What most of us today call the "rainbow trout" became the newest entrant in the Commission's experiment.[25] Two important questions needed to be answered: Into how many areas of the country could the Commission successfully introduce this other cold-water fish? Equally important: Would the public favorably take to this new fish, both as a food source and as an angler's prize?

Stone and two assistants left the Baird salmon station in late July 1879 to search for a proper site to harvest trout. Rainbow trout were abundant in the McCloud, so finding the best location to harvest their eggs really boiled down to convenience. About four miles up river, they found a suitable spot with a cold-water feeder creek that ran into the river. Because it had all the essential characteristics (e.g., constant flow of cold spring water, level terrain), Stone and his crew soon set about to construct a hatching house and dig holding ponds.

Collecting rainbow trout eggs used a different approach than that for harvesting salmon eggs. Instead of netting or trapping them like salmon, rainbows were caught using set lines: cords 150 to 175 feet in length, with finer drop lines spaced about every 5 feet. A hook was tied to the bottom of these drop lines. Ironically, salmon eggs were the most successful bait. Breeder trout were caught using these set lines and transferred to the holding ponds. Once the females were ripe, the eggs were harvested and fertilized using the same methods as with salmon. Once fertilized the ova were placed in trays and aged in the hatching house. When the eggs were ready for shipment, boxes containing rainbow trout eggs were packed into crates and the crates were shipped to applicants in the same manner as the salmon eggs. Because the methods used to ship the trout eggs mimic those used to ship salmon eggs, I need not repeat that discussion here. Obviously, the addition of trout eggs being shipped expanded the Commission's dependence on the railroads and their employees.

The first rainbow trout eggs were harvested in early January 1880. By the end of May, a total of about 388,000 rainbow trout eggs had been collected, of which 261,000 had been shipped by rail to hatcheries back east. Of the eggs taken, Stone hatched 68,000 fry and returned them to the McCloud.[26] Stone's records show that the first shipment of rainbow trout eggs that left the McCloud on 30 January 1880 were destined for the Druid Hills hatchery in Baltimore. Maryland's fish commissioner, T. B. Ferguson, reported that the fry hatched from these eggs were released on 15 April 1880 into a tributary of Gwynn's Falls.[27] Maryland was the first *government* hatchery to receive rainbow eggs from Stone and release them in this new program. They probably were not, however, the first hatchery in the East to receive a shipment of trout from California. Though there is some debate, that claim belongs to none other than Seth Green, who is credited with establishing the first brood stock of rainbows outside of their native range.[28]

With the trout operation up and running, Stone was soon shipping hundreds of thousands of rainbow eggs to federal, state, and even private hatcheries around the country. Unlike the salmon experiment, the hatcheries to which the rainbow eggs were sent were much more geographically dispersed. This reflects the fact that, as more and more rainbow fry were being stocked around the country, the evidence indicated that this species was not only robust to its new environments, but also quite popular with the locals. Wakeman Holberton, a New York artist known for his still life, wildlife, and landscape paintings, wrote to Commissioner Baird that while visiting Ohio in 1882 he found the "California trout that we put in [a stream near Cleveland] in 1881 were doing finely last year and had already grown to the size of four inches." Roland Redmond of the South Side Club, an association of fishing enthusiasts in New York City, requested from Baird an *additional* shipment of rainbow eggs in 1883. The club's initial shipment of eggs in 1880 had been hatched and stocked, and some had already reached twenty-two inches and weighed three pounds. Needless to say, the club was "anxious to stock one of its ponds with this [rainbow trout] fish."[29] I could repeat such testimonials many times over—Baird made sure to publish them in the Commission's annual *Bulletin*—but I think you get the idea: those who received a shipment of rainbow trout and stocked them in local waters often found that they not only survived but flourished.

The public soon became enamored with the rainbow. So did state fish commissions. State hatcheries receiving rainbow eggs from Stone discovered that they were easy to propagate. The Nebraska Fish Commission's request for another shipment of rainbow eggs is illustrative. "With one other lot of 10,000 *we hope to establish ourselves securely in the production of all we need for future operations.*"[30] This request is important. The Nebraska Fish Commission, like many others across the country, found that with a starter batch of eggs from Stone they could establish their own brood stock. Once this was done, they could almost perpetually extract eggs, fertilize them, and stock fry into their local streams and ponds. The rainbow was a dream come true for the commissions, both state and federal: they had found a fish that could be easily propagated, was a fine food fish, and was considered by anglers a highly prized sport fish.

The success with establishing brood stocks outside of Northern California spelled trouble for Stone's hatchery in California, however. Not only were hatcheries run by state commissions (and private individuals) producing trout for their own use, but so, too, was the Commission itself. Marshall McDonald, then the chief assistant commissioner to Baird, reported on the Commission's distribution of various species of fish during 1885 and 1886. By this time a large number of the rainbow trout eggs being distributed to state and private hatcheries were coming not from California but from the federal fish hatcheries in Northville, Michigan, and Wytheville, Virginia.[31]

The ease with which rainbow trout could be propagated made it a perfect fish for the wide dispersal sought by the Commission, still operating under the parameters of the joint resolution. The Commission (and many states) was engaged in a broad effort to stock rainbow across states where they believed they could seed the waters and establish perpetuating wild populations. The idea was the same as with salmon: stock rainbow fry in streams where they would grow and, hopefully, reproduce to create a sustaining population of fish that could serve as a food source. Though wildly more successful than salmon plantings, the success of having the rainbow fry reach adulthood was mixed. One solution, of course, was simply to plant more and more fry. But there was a better way.

McDonald recognized that trying to create self-sustaining rainbow populations in streams across the country proved to be "disappointing and wholly incommensurate to the expenditure incurred."[32] The Commission's rainbow trout program thus took a significant turn in 1886. McDonald's plan was to propagate the trout in the Commission's increasing number of hatcheries and distribute the resulting fish in streams across the country. But, instead of depositing fry, the Commission would shift to transporting and planting rainbow that were four to six inches in length. The idea was that by planting these larger fish, they would be more likely to survive. Of course, this decision made Stone's work on the McCloud station redundant. By this time his salmon operation had already been significantly scaled back, and now the trout operation was being closed down.[33] Stone's trout-taking continued, though, on a much smaller scale until early 1887. On 16 April the last shipment of rainbow trout eggs was sent to the federal fish hatchery in Washington DC. McDonald's change in policy reflects a change in the thinking about how best to stock fish. It also reflects the fact that the Commission had adopted a new technology to ship fish: the dedicated fish car. That will be the focus of the next chapter.

Just how many rainbows were shipped? The Commission distributed a total of nearly 286,000 eggs, fry and adult rainbows in 1885. Five years later, after the McCloud hatchery had been closed, the Commission sent 140,000 eggs and almost 55,000 adult fish to state hatcheries and individuals across the country. It also engaged in stocking trout, so these distributional numbers probably undercount the true quantities And they ignore the fact that many states and individuals were propagating and stocking rainbows, so the total size of the rainbow trout program is understated by the Commission numbers alone. By the end of the 1800s the Commission was annually distributing almost 367,000 eggs, fry, and larger fish to applicants around the country. And even though these numbers are low compared to the number of shad being shipped (over 235 *million*), the Commission, its state counterparts, and private individuals would continue to propagate and stock rainbow trout for years to come. Indeed, although the shad and salmon experiments did not successfully meet their original goals, to this day the rainbow trout is abundantly found in nearly every state. The rainbow trout experiment was so successful that Anders Halverson has wryly observed that the rainbow trout beguiled America and overran the world.[34]

Earlier I suggested that there were two fish that helped restore the Commission's reputation following the disaster of the salmon experiment. The exotic rainbow trout captured the hearts and minds of fish culturists and anglers across the country and even the world. The other fish that gave the Commission's standing a boost amongst the public and the politicians? The very nonexotic carp.

Carp

The story of introducing carp to America is just not as sexy as salmon or trout. There wasn't the allure of a prized fish being transplanted from one side of the country to the other and everywhere in between. There are no tales of setting up hatcheries in the wilds of California. And, unlike salmon and trout, the carp did not appear on most anglers' wish list. Like shad, carp were introduced to serve as a food fish. As such, one aspect of the carp story that is a bit different is that it did not begin with a fish commission—federal or state—experimenting with its propagation and introduction. Instead, in the late 1860s and early 1870s, private fish culturists were importing carp to see if they could raise it as a marketable food fish.[35]

Unlike many of the other fish with which the Commission experimented, carp thrived in almost any type of fresh water. This attribute meant that carp could be distributed in many more states than, say, trout. Carp were important for states where the public's access to protein sources was dwindling. Bison were being eliminated by professional hunters; the population of other wildlife like deer, quail, and turkey also were becoming increasingly scarce. Politicians looking for a remedy to the exploitation of the native wildlife recognized that "the carp fitted the picture as a useful, 'civilized' animal which could be stocked and replenished easily, and which could become a source of cheap protein when native game was increasingly in short supply."[36]

Baird and other, but not all, fish culturists knew of the carp's fecundity and its ability to thrive even in poor water conditions. (Think muddy ditches.) Baird noted the carp's popularity in Germany "for church requirements and the general wants of the table" and "for a long time attached much more importance to the introduction of carp into the United States of America," especially to "supply an often-expressed want of fish for the South."[37] Carp thus enabled the Commission to fit its mission as prescribed in the joint resolution: increase the nation's stock of food fish.

The beginnings of the Commission's carp experiment were inauspicious. The German fish culturist Rudolph Hessel brought carp to the United States from Germany in 1876, though only a few survived the journey.[38] Hessel tried again in the spring of 1877, this time bringing 345 carp to New York, eventually moving them to the hatchery ponds near the Washington Monument in Washington DC a year later. To grasp the potential that carp offered, from these initial fish the Commission was able to produce nearly 4,800 carp fry that were shipped to fifteen individuals in thirteen states only two years later.

Commissioner Baird fueled the public's interest in and demand for carp by hyping its reproductive and growth capabilities. Carp represented the perfect food fish: it could be stocked into almost any water system—rivers, lakes, small farm ponds—and it would thrive. Unlike salmon or trout, it seemed impervious to its surroundings. Carp fever spread like a pandemic.

By 1883 the Commission labored to meet the demands of over 5,500 applicants—mostly individual landowners—throughout the country. Because of production constraints, the Commission shipped only 144,000 carp fry to applicants: about 1,200 orders went unfilled.[39] The Commission and the few states that already had established their own brood stocks were providing carp fry to just about anyone who requested them. The supply simply could not keep pace with demand.

As you have discovered by now, the Commission seemed oblivious to any negative environmental effects stemming from their actions. Carp, by their very nature, are aggressors. They eat native fish, disrupt balanced ecosystems with their feeding habits, and strain food sources by the very fact that they are so prolific. The Commission's myopic mindset was to transplant fish, including carp, wherever they could. One example of this attitude is Yellowstone Park Superintendent Philetus W. Norris's desire to introduce carp into the park in 1881. Luckily for future generations, park administrators terminated the introduction of any non-native species into the Yellowstone's waters.[40]

The Commission's widespread distribution of carp served two purposes. First, it affirmed that the Commission could deliver on its promise to restock the nation's fishery. One observer captured the interest in carp by noting that "public discourse was initially dominated by carp advocates led by... Spencer Baird, whose hype about its potential as a food fish for farm ponds sparked a 'fever of enthusiasm' for the thousands of free carp distributed by the [Commission]."[41] Second, it used early shipments of carp as a test case for its newest technological improvement in the application of fish culture: the fish car.

By the mid-1880s the Commission was shipping almost 350,000 carp fry to private individuals and state commissions for deposit in public waters. Many state commissions jumped aboard the carp bandwagon and began propagating and distributing carp on their own.[42] The carp became so popular that the

Missouri State Fish Commission abandoned its propagation of salmon and trout at its cold-water hatchery in St. Joseph in 1884 and dedicated the facility to the production of carp. One reason was economic: carp were more easily and cheaply produced. The other was political: because they could grow anywhere, carp were more popular with the public and, therefore, the politicians who voted on the commissions' budgets.

Before too long the curtain was drawn back, and the downside of stocking carp here, there, and everywhere became all too evident. Carp were overrunning their adopted surroundings, chilling the public's love affair with them. Disillusion and even a growing aversion to the carp is illustrated by this statement by the Missouri State Fish Commission: "We would warn you against the so-called German Carp," it writes in one of its annual reports. But it goes on, wondering, "What facility ever induced the introduction of this contemptible and filthy and filth-loving fish into the waters of North America, this Commission cannot say, for aside from its rapid growth [and] fecundity, it has nothing to recommend it to the sportsman or the fish farmer."[43] And this from a commission that less than a decade earlier had published the pro-carp pamphlet *Carp and Carp Culture in Missouri with Appendix on Native Fish*, which extolled the virtues of the carp, praising its taste and the ease of raising it.

The Commission abandoned its carp distribution program in 1896. Many state commissions followed suit. How quickly the public soured on the carp. In the span of about twenty-five years "although carp remained unchanged biologically, it evolved rhetorically from pond fish to savior of commercial fishing, from esteemed delicacy to muddy-tasting trash fish, from foreigner to native, from welcome addition to unwanted pest."[44] Unfortunately, the carp's widespread effect on the nation's waterways continues to this day. Pandora's box had been opened, and there was no putting the evils back.

In Summary

The major focus of the Commission's freshwater activities during its initial years was distributing eastern shad and Pacific salmon across the country in order to try and establish self-sustaining populations. The shad experiment worked to some extent: it appears to have helped stem the downward trend of populations in the East and established a foothold in the West. No shad migrations ever developed in the midsection of the country, however. The salmon experiment was a complete bust, even in the eyes of its strongest advocates.

When the Commission added rainbow trout from Northern California and German carp to the list of potential transplants, they discovered two species that were highly adaptable and proved themselves as suitable candidates for its program of rebuilding the nation's fish stock. Of course, history tells us that the widespread stocking of each had their own set of effects on their new ecosystems.

I do not want to leave you with the impression that shad, salmon, trout, and carp were the only species that the Commission experimented with. By the end of the 1800s the Commission and state commissions were transplanting dozens of different species of freshwater fish around the country, to the tune of over one billion eggs, fry, and adults of the various species being shipped in 1899 alone. Some were deposited into limited regions—tens of millions of whitefish were deposited into only the Great Lakes. Others found new homes based on the Commission's approach of putting fish wherever they wanted, or gave them to whoever requested them.

To put this into context, table 1 shows the history of fish introductions by the Commission to just one state: California. The table includes not only the species of fish, but also the state(s) where they originated, when they were introduced, and their current status.

Table 1. Fish introduced into California, 1871–98

Year	Species	Source	Attempts	No.	Status
1871	eastern brook trout	NY; WI; NH	12	1,530,000	Abundant
1871	American shad	CT; NJ	7	834,000	Abundant
1872	whitefish	MI	6	1,6000,000	Failure
1874	Atlantic salmon	ME	1	305	Failure
1874	catfish (3 species)	VT; NE; NJ	1	344	Abundant
1874	black bass (2 species)	VT; NE?	3	1,577	Abundant
1874	glass-eyed perch	VT	1	16	Failure
1874	silver eels	NY	3	4,022	Failure
1874	rock bass	VT	1	4	Failure
1877	carp	Germ.; Japan	3	386	Abundant
1878	land-locked salmon	ME	5	96,000	Failure
1879	striped bass	NJ	2	450	Abundant
1891	green sunfish	IL	1	250	Abundant

Year	Species	Source	Attempts	No.	Status
1893	muskellunge	NY	1	93,000	Failure
1895	brown trout	Germ.; Scotland	3	323,000	Abundant
1895	lake trout	MI?	2	65,300	Present
1896	pickerel	?	1	27	Failure
1896	yellow perch	VT?	1	454	Present
1898	golden shiner	eastern U.S.	2	2,253	Abundant

This table shows the variety of fish species that were introduced into California from 1871 to 1898. There are nineteen different species, about half of which are still found in the state.

Source: Jerry C. Towle, "Authored Ecosystems: Livingston Stone and the Transformation of California Fisheries," Environmental History 5, no. 1 (January 2000): 55.

Between 1871 and 1898 nineteen different species of fish were introduced into California waters. Two species originated from outside the United States: carp from Germany and Japan, and brown trout that were being imported from Scotland and Germany.[45] The rest were natives, mostly from the eastern United States. This latter fact reflects the degree to which American fish culturists were trying to make over fish populations in the West. It also shows the habit of humans to move familiar items with them as they migrate, bringing the flora and fauna from their homelands to new surroundings.[46] Another takeaway is just how experimental (haphazard?) the experiment really was. Against the nine "success stories" in the table there are eight failures, though in some cases the number of attempts and the number of fish planted did not auger a very high probability of success. In some cases, such as the pickerel, it is good that it failed to take hold. A number of stories in publications like *Forest and Stream* reported that where they had been planted in the East they were highly detrimental to native species. And, of course, the success of the carp remains a point of debate to this day.[47]

The Role of the Railroads

The Commission's experiment to transplant millions of fish eggs and fish across the country was made possible only through the assistance of the railroads. And I do not mean the mere existence of the railroad network. The railroad system was not a public good like modern roadways, but a transportation grid owned by private companies. Yes, the railroads sometimes charged the fish commissions for shipping egg crates or the use of their track when commissions began using fish cars. But we know from comments in the Commission's annual reports that the rates usually were far lower than that paid by the public. And, more often than not, fish messengers and their cans of fish rode for free. The reports of stocking trips also indicate that railroad employees assisted the messengers in moving their egg crates and milk cans and sometimes were the ones who actually made the deposit in a nearby river. From the outset such assistance to make this grand experiment achievable did not go unappreciated.

Baird's *Report of the Commissioner for 1873–1874* began a tradition that carried on through the rest of the century: public recognition of the railroads that provided support to the Commission. In this *Report* Baird lists ten railroads on which "the cans [of shad fry] were carried as baggage, without extra charge, and access to them afforded at all times."[48] For these ten railroads some individual—a depot master, general transportation agent, or even the company's superintendent—with whom the shipping arrangements had been made was called out. Baird also acknowledged the help of six other railroads, but separated them out apparently because they levied a baggage fee for the fish cans. Pretty good PR for those companies that helped in accomplishing the Commission's grand plan.

What is noteworthy about Baird's *Report* covering Commission activities for 1872 and 1873 is that he makes public his correspondence with the railroad and express company executives regarding how they were assisting the Commission. The correspondence suggests that when Baird sought help the companies were more than willing to oblige. (Of course, those who did not were not likely to be found in the report!) A good example is William G. Fargo's message to his American Express Company employees. Fargo informed them that they should "aid in every possible way the parties [fish messengers] who are engaged in the transportation of live fish, with which to stock the western waters."[49] A. J. Cassett, the general manager of the Pennsylvania Railroad Company, similarly informed his "baggage-masters [to] allow agents of the Commission . . . to ride in the baggage-cars, when it is necessary for them to do so. . . . The tanks [fish cans] will be carried free of charge."[50] Such memos from the home office to railroad employees could be repeated severalfold. They show that from the outset railroads became integral in seeing that the Commission's experiment in managing the nation's fishery was conducted in a manner that gave it the best chance of succeeding. And if you helped the Commission, the Commissioner would let the public know about it.

Although I can find no extant record to prove this, I am sure that Baird and the railroad's lobbyists kept Congressmen well apprised of the railroads' public service. You might recall from chapter 1 that by this time the railroad industry was notorious for its bad behavior. In light of their often noncompetitive actions, an obvious question to ask is, "Why would a profit-seeking private company provide its service at a reduced cost or for free to some government agency?" As the example of Oregon's senators negotiating with the Commission to manage its salmon hatcheries shows, sometimes political pressure was brought to bear on the railroads to assist the Commission. But I think the majority of cases represent something else. As the correspondence with Baird indicates, some railroads simply considered their help to the Commission as the public-spiritedness of seeing this grand experiment succeed. A letter to Baird from Samuel M. Shoemaker, resident manager for the Adams Express Company, states that by "securing prompt transmission of [the cans and the messengers] and the procuring of fresh water along the route during the stopping of the trains [the company] *will be doing a public service*" (emphasis added).[51] The Commission's stocking of fish around the country was quite popular, and any railroad involved could ride the Commission's coattails of public appreciation.

Some companies capitalized on the public relations opportunities afforded by aiding the Commission. Henry Sanford of the Adams Express Company told his stationmasters to make sure that "at points where the Commission leaves spawn for deposit in the rivers or other streams of the vicinity, [you] *will request the aid of the local authorities, and invite them to witness that the spawn is properly deposited at the earliest possible moment after reception*" (emphasis added).[52] The arrival of a Commission's fish messenger and his fish were publicity events to which the local papers and city leaders should be invited, and the railroad might as well piggyback on the popularity. Think of the headline: "U.S. Fish Commission Stocks Local River with Free Fish. Adams Express Company Makes It Possible."

The list of railroads who had "granted facilities" to the Commission increased to forty-one companies in the 1877 *Report*. Baird even referred to several companies for special mention, heaping praise on the Baltimore & Ohio Railroad; the Philadelphia, Wilmington & Baltimore Railroad; and the Pennsylvania Railroad. All of these companies allowed the Commission's fish messengers free access to their cargo, most often located in the baggage car of express trains.

As usual Baird made clear that these railroads charged no baggage fee for the cans of fish, and that the fish messengers traveled for free.

The Baltimore & Ohio Railroad got a special nod in Baird's 1877 *Report*. The reason is because the company's management authorized "the stopping of trains at such points on rivers or streams as it might be convenient for the proper introduction of fish into the waters."[53] Not only were railroads transporting the Commission's fish eggs and cans of fish, but many, like the B&O, were also stopping their trains wherever fish messengers (or the railroad employees) wanted to dump a few cans of fry into a nearby stream. Such behavior was not uncommon. For example, the stocking records for Missouri often report the location of a deposit site simply as "along the Frisco track," the "Frisco" being the nickname for the St. Louis–San Francisco Railroad, which ran across the state. If railroad companies hauling fish and fish messengers for little or no cost were also willing to interrupt their schedules to stop at this or that stream crossing to aid the stocking of fish, I wonder if the passengers thought such delays made them part of the grand experiment? In some instances passengers became willing aides to the messengers, so I guess they, too, got in the spirit of what the Commission was up to.

Railroads also acted as agents for the Commission when it came to transferring fry to applicants. It was standard practice for a fish messenger to telegraph ahead (often with a notice of a day or two) and let recipients know on which train he and the applicant's fish were traveling, at which station they could acquire their fish, and what time they should be there. When the train arrived at the predetermined station, the fry were transferred from the Commission's milk cans into whatever containers the applicant brought. If applicants failed to bring suitable containers, they were allowed to borrow the Commission's fish cans, but only if they ensured their return to the depot. In these instances the railroad's stationmaster assumed responsibility for making sure that the cans got back to the Commission.

But missed connections were bound to occur: recall the difficulties in matching railroad schedules across different train lines? If it became clear that the messenger and the fish would not arrive as scheduled, an applicant might receive a telegram stating that, due to the unforeseen difficulty—perhaps the fry were stressed—the fish had been deposited in some stream on the way. In these cases applicants could reapply for another delivery of fish. If applicants were late or did not show up, railroad employees often were left with the task of depositing the fish in any nearby stream or lake of their choosing.

Federal and state fish commissions deeply appreciated these services provided by the railroads and recognized that doing so was an imposition. A wonderful example is the front-page message that appeared in the 2 February 1876 *Hillsdale Standard* in Hillsdale, Michigan. Written by George H. Jerome, Michigan's superintendent of state fisheries, the article explains how private individuals interested in stocking their lake with whitefish should go about applying to the Michigan commission. It includes a laundry list of the information needed to request the fish, including the recipient's name, address, name of lake, and depth of water. The request also required the name of the railway station at which the applicant would call for their fish because the "commissioner can deliver no fish except at railway stations." Because the railroads were involved, Jerome informed the "depositors," as he referred to applicants, that when the train stopped at the predetermined station they must quickly present themselves at the railroad car's baggage door to "assist in removing the cans [of fish] from the car and *to occasion as little delay to the train, and trouble to the employees of the road as possible*" (emphasis added). After all, Jerome intoned, it is because "the [rail] roads of the state are very kindly moving our fish free of charge, and hence every consideration of honesty and fairness requires that we should facilitate the transit of the cans by every possible means and agency at our command."[54] Superintendent Jerome understood the importance of the railroads in the state's plans for restocking and he wanted to make sure the public knew it, too.

Complications in the transfer of fish from fish messenger to recipient were bound to occur. Even so railroads often went out of their way to make the fish stocking a success. The instructions communicated from the superintendent of a Maryland rail line to its station masters are illustrative.

> Several shipments of live fish will be made by the Commissioner of Fisheries to parties in the neighborhood of your station. In delivering them to the consignee, you will in no instance issue the vessels in which the fish are sent. The parties receiving the fish must provide vessels for their removal to the ponds.
>
> A blank receipt will be forwarded with each shipment, and you are required to have the same filled up and signed by the consignee, and return the receipt to the Commissioner of Fisheries.

Should the fish not be promptly called for, you are requested by the Commissioner to change the water in the cans once or twice a day, or whenever the fish show signs of distress by coming up and remaining at the top of the water. If the fish are not delivered within 5 days after their arrival, you will report the fact to this office and await instructions. After delivering the fish return the empty cans as soon as possible to this office, regularly way-billed FREE.[55]

State fish commissions like Maryland were engaged in stocking and relocating fish and also relied heavily on railroads. The note illustrates that they also would extend to railroads the same courtesies as did the Commission. And, like the Commission, the states publicly acknowledged the railroads in their annual reports. The 1876 annual report of the Commissioners of Fisheries for Maryland not only included a broad thank-you to the railroads involved but explicitly named names. Special notice went to Vice Presidents John King and William Keyser of the Baltimore & Ohio Railroad and ex–Vice President DuBarry of the Baltimore & Potomac Railroad for providing free transportation. President J. H. Hood of the Western Maryland Railroad and Mr. George Wilkins of the Northern Central Railroad also were mentioned for providing free transportation and because of their "general orders, which insured the assistance of their agents."[56] One annual report of the New York Commissioners of Fish, Game, and Forests noted that "nearly every railroad in the State has rendered this aid freely when called upon so to do; but we are especially indebted to the Delaware and Hudson Canal Company, the New York, Ontario and Western Railroad, the Delaware, Lackawanna and Western Railroad, the Buffalo, Rochester and Pittsburgh Railroad, and the New York Central and Hudson River Railroad, as they are more nearly connected with the hatcheries of the State."[57]

The common theme is free transportation, but railroads provided more than that. The 1877 annual report from the Missouri Fish Commission makes the astute observation that "this generous charity [by the railroads] has enabled us to begin stocking waters with valuable fish. *Without this aid, we could have done nothing in this direction*" (emphasis added). Of course, "all the railroad companies in Missouri, without exception, when called upon, have given us free transportation" for the fish messengers and their milk cans.[58] Perhaps it was obvious, but Missouri's report exposes the reality behind the Commission-railroad partnership: the experiment to relocate and stock fish in such massive numbers at the state and national level simply could not have occurred without the railroads' involvement.

Why Did Railroads Provide Assistance?

I have been unable to find any direct mention of a company's involvement in the Commission's fish relocation experiment. Some may have done so, but my admittedly nonexhaustive search of numerous railroad annual reports did not turn any up. Most newspaper articles about the Commission's stocking activities mention that it took place near some line's track, or that the fish were delivered by this or that railroad company. So, without firm evidence, other than the quotations provided earlier—doing a public good, garnering some public recognition in helping with the stocking—why *did* railroads help the fish commissions?

One possible explanation is that they were patriotically supporting an experiment with important national implications. The experiment captured public interest, not unlike the building of the transcontinental railroad. Local newspapers carried stories, sometimes reprinted from larger outlets, about "fish culture" and the national importance of the fish stocking program. Recall the story about Seth Green's work that found its way into the *Clinton Advocate* in western Missouri. Because the availability of fish as a food source was a national concern, it is perhaps not surprising that railroads offered their help out of some sense of public spirit.

Another explanation is that railroad companies, especially during the latter decades of the century, were not adverse to some positive press. If the public were made aware of the fact that the railroads were providing their services to the Commission for free (or at substantially reduced rates), that trains stopped to offload a can of fry into nearby streams, or that station agents assisted the fish messengers, such positive press could only improve their image. You might recall that the railroads were not the darlings of commerce that they had once been, coming under attack for a variety of (valid) reasons. It is not a stretch, therefore, to suggest that by cooperating with the federal and state fish commissions in such a popular enterprise they could, at a very low price, acquire some positive public relations. And surely those in Congress and state legislatures would recognize such public-spiritedness when it came time to vote on bills affecting the industry.

Lastly, as a preview to arguments made in future chapters, railroads were businesses and therefore always looking for ways to increase their revenues. Passenger travel represented a growing source of revenue. Pleasure travel was on the rise as more in the population looked to escape the city and explore the outdoors. By assisting commissions in the stocking of fish, railroads were helping to maintain or even improve the fishery to which it could provide the best access. In helping the fish commissions introduce exotic species like trout into new waters, railroads also were helping to establish a pastime that catered to the increasing number of individuals—both men and women—seeking to try recreational fishing. By helping the commissions, railroads were helping to create a new clientele of outdoorsmen and women that would pay to use their service.

A Final Note

The Commission actively transplanted fish from one location to another in an attempt to replenish the nation's fishery. It started with transplanting eastern shad and Pacific salmon from one coast to the other and to points in between. Rainbow trout from Northern California and carp originally from Germany soon were added to the transplant list and stocked in waters across the country. Over time this activity expanded to include dozens of species being stocked by the Commission (and their state counterparts). And so did the geographical reach of these fish-stocking activities.

The scope of this undertaking simply would not have been possible were it not for the railroads. These private firms shipped egg crates, milk cans of fry, and the fish messengers who attended them for free or at below-market rates. For their part railroads were key partners with fish culturists in the various fish commissions in shaping the future of the nation's fishery. As critical as railroads were, this partnership went to another level once the commissions adopted the newest technology in fish culture—the fish car.

5

THE FISH CAR

Relying on fish messengers tending to several cans of fish was inadequate if the Commission wanted to ramp up its scale of operations. Two issues arose: first, because a single fish messenger could handle no more than ten or twelve cans, this limited the number of fish per delivery. Second, a large percentage of the fry were lost to predation once they were deposited. Even if a messenger stocked, say, one hundred thousand fry into some stream, a very small percentage would ever reach adulthood and reproduce. The only way to address these issues was to figure out a way to transport more and larger fish than possible using fish messengers with milk cans. The answer was the fish car.

The story of the fish car is important because it reflects the Commission's eagerness to expand the scope of its work. The fish car represented a significant advancement in fish culture, one that depended on innovations in railroad technology. With its fleet of fish cars, making large-scale deposits of larger fish across a broader geographic area became a reality. Like most stories, it is best to begin at the beginning.

The California Aquarium Car

The possibilities of stocking non-native species in local lakes and rivers captivated the fish culture community, both inside and outside the Commission. The California Fish Commission was one such group. Made up of East Coast transplants, members of the California commission wanted to join the experiment to transplant fish from their roots to the waters of their new home. As I've already mentioned, this meant sending not only shad from the East Coast, but also California trout to fish culturists in the East. Doing so milk can by milk can seemed inefficient. So, in the spring of 1873, the California commission purchased a fruit car from the Central Pacific Railroad Company. The car was specifically designed to make rapid trips across country, fruit being subject to spoilage. The size and weight of the fruit car allowed it to travel with passenger trains instead of being relegated to slower freight trains. The car was equipped with the recently developed Westinghouse air brake system and a Miller platform, a relatively new invention that, in the event of a collision, prevented the train cars from telescoping together.

The California commission transformed the car from one carrying fruit to fish. An eight-by-eight-foot tank capable of holding over 1,200 gallons of water (about five tons) was installed at one end of the car. An ice box was fitted at the other end. Given the description of moving fish by milk can found in the last chapter, you can appreciate what this innovation meant. Attending fish messengers (and the train's crew) now had water and ice readily available for keeping the water in the milk cans refreshed and cool. A framework to hold the milk cans was built along each side of a middle aisle. This prevented them from moving about during transit. Bunks in which the messengers could sleep folded down from the ceiling above the array of fish cans. Compared to schlepping fish cans between baggage cars numerous times in even fairly short deliveries, the fish car represented a significant improvement. Not only could more fish be transported in one trip, but the messengers attending them also had ready supplies of ice and water, not to mention basic living quarters.

The California fish car was first used in a joint project between the California and U.S. fish commissions to relocate a variety of fish and aquatic species found in eastern waters to California. Some deposits also would be made in some western states along the way. Livingston Stone was selected to oversee the project. In March 1873 Stone left the salmon hatchery on the McCloud River and preceded the California fish car—also known as the aquarium car—back to his home of Charlestown, New Hampshire. Once he arrived he collected the various species of fish and other creatures requested by the California commission. It was quite a menagerie: black bass, wall-eyed pike, glass-eyed perch, yellow perch, bullheads, tautogs, saltwater eels, brook trout, shad, lobsters, and oysters. Most came from the East, though a few species on the wish list (e.g., catfish) would be acquired en route.

As soon as the car arrived in Charlestown it was made ready for the trip back west: loaded with water, ice, and milk cans containing the assorted aquatic species. With the loading completed, the cargo and a crew of four—Stone, W. T. Perrin, Myron Green, and Edward Osgood—set out from Charlestown on 3 June to begin their journey back to California. No one had yet collected such a wide variety of aquatic species for such a long trip. This maiden voyage of the fish car would help the Commission determine if these various species could survive the journey and, over time, their new environs. It therefore represented a milestone in the budding experiment to combine fish culture and rail travel.

Stone and the fish car made their way westward with stops in Albany and Detroit. At these stops the car was transferred, respectively, from the New York Central to the Great Western to the Michigan Central Railroads. From Detroit the car headed to Chicago, where it was switched over to the Chicago & Northwestern Railroad bound for Omaha. In Omaha it would once again be transferred to the Union Pacific, which would carry it all the way to California. Along the way the crew enjoyed the courtesies of the different railroad companies. In Omaha, for instance, the fish car was shuttled by the Union Pacific to its own icehouse, where more than a ton of ice was loaded for free.

Stone and the California fish car reached Omaha on Sunday morning, 8 June. Only five days and about halfway into its inaugural trip, the car was attached to a Union Pacific train bound for California. The Union Pacific train departed from Omaha in the early afternoon. With the ice box restocked, the water tank filled, and the water in the fish cans refreshed and aerated, Stone and his crew settled in for a quiet lunch. Or so they thought. About twenty miles west of Omaha, the train crossed a trestle over the Elkhorn River. Normally a muddy and slow-moving river, the Elkhorn had overflowed its banks from recent rains. Unbeknownst to the Union Pacific and the engineer, the trestle spanning the river had been weakened by the flooding. As the train crossed the trestle it collapsed. Let Stone describe the unfolding state of affairs:

> We made our tea and were sitting down to dinner, when suddenly there came a terrible crash, and tanks, ice, and everything in the car seemed to strike us in every direction. We were, every one of us, at once wedged in by the heavy weights upon us, so that we could not move or stir. A moment after the car began to fill rapidly with water, the heavy weights upon us began to loosen, and, in some unaccountable way, we were washed out into the river. Swimming around our car, we climbed up on one end of it which was still out of water, and looked around to see where we were. We found our car detached from the train, and nearly all under water, both couplings having parted. The tender was out of sight, and the upper end of our car resting on it. The engine was three-fourths under water, and one man in the engine-cab crushed to death. Two men were

floating down the swift current in a drowning condition, and the balance of the train still stood on the track, with the forward car within a very few inches of the water's edge.

The maiden voyage of the "celebrated California aquarium car" ended in tragedy. Not only was there loss of human life but "one look," Stone recalled, "was sufficient to show that the contents of the aquarium car were a total loss. No care or labor had been spared in bringing the fish to this point, and now, almost on the verge of success, everything was lost."[1]

The *Omaha Herald* reported the accident under the headline

> RAILROAD ACCIDENT
> An Engine and One Car Breaks Through
> the Trestle-Work, near the
> Elkhorn on the Union Pacific
> Roadmaster Carey Killed—Narrow
> Escape of Others
> Wreck of the Celebrated California
> Aquarium Car
> Personal Experiences of the Survivors[2]

You'd expect the local press to cover the story, but the news that the California fish car had been destroyed garnered wide attention. This, I think, is evidence of the general public's interest in the Commission's fish-stocking activities. For example, the *Perrysburg Journal* in Wood County, Ohio, ran a front-page story about the accident. After describing the facts of the derailment, it included the droll observation that "it [the Elkhorn] is said to be the best stocked stream in Nebraska at present." Indeed, soon after the accident stories circulated, a few enterprising locals gathered up whatever fish they could and deposited them into nearby ponds.

Such a near-death experience might give one pause, but Stone seemed undeterred. The day after the accident he telegraphed Baird in Washington DC and Mr. Throckmorton of the California Fish Commission to inform them of the accident and to await further instructions. Baird told him to return east with his crew to oversee a shipment of shad from the Commission to California, the journey that I detailed in the previous chapter. There is no record of Throckmorton's response.

The following year Stone would oversee another coast-to-coast shipment of fish to California, this time in the second [not celebrated?] California aquarium car. What is interesting, perhaps informative, is that Stone's report of his second trip is, well, terse. His account of the inaugural 1873 trip runs some six pages in the Commissioner's *Report for 1873–74 and 1874–75*; the 1874 trip garners only two pages, with a paltry six lines of text. A table listing the species carried, including where they were procured and where they were deposited, and the fish car's itinerary fills most of the report. In a nutshell, the second California aquarium car left Charlestown on 4 June 1874 with an aquatic passenger list that varied only slightly from the first: no shad and trout this time but now including Penobscot salmon destined for the Sacramento River. The second trip took eight days. The fish car arrived safely in San Francisco the night of 12 June. Some of the fish had been deposited along the way, including oysters and a few lobsters dropped into the Great Salt Lake. In a somewhat ironic twist, local catfish were collected from the Elkhorn River in Nebraska, destined for the San Joaquin River in California.

Perhaps the absence of a major derailment lessened the noteworthiness of the second trip. Or, perhaps, the new had quickly become mundane. If Stone thought the second trip unremarkable, the public did not share this opinion, however. The Rochester *Democrat* on 6 June ran a story about the trip of the second fish car and its purpose and noted that it had passed through town on its journey westward. The article also listed the various species being transported and even gave the measurements for the car itself. The *Omaha Bee* ran a similar story about the car's passage through town, its objective, and so on. Such coverage probably would not be too surprising if the fish car traveled through your town. But, as we've already seen, the public's general interest was wider than that. Why else would the newspaper in Milan, Tennessee, a town not on the fish car's itinerary, think it newsworthy enough to reprint the Rochester *Democrat*'s article?

Stone's arrival in California proved the viability of a railroad car specifically designed for the long distance transport of a large number of varied fish species. This innovation, not only in this aspect of fish culture, but also in the growing ease with which transcontinental travel occurred revolutionized how the Commission could approach its management of the nation's fishery. What is surprising is how long it would take the Commission to adopt this new technology itself.

The Commission's First Fish Car

Throughout the 1870s railroad companies lined up to assist the Commission and apparently any of the newly forming state fish commissions that asked for help in their fish-transplanting activities. The commissioner's *Report for 1880* lists 113 railroad companies actively working with the Commission (no work for state commissions is listed) in one form or another. This is quite a jump from the 42 railroads just a few years earlier. Greater involvement by the railroads meant that the Commission could extend its reach over an ever-broader geographical area. Using the home office of the listed railroad company as my guide, the railroads in Baird's 1880 *Report* cover thirty different states. Since many of these railroads operated across state boundaries, it is likely that all thirty-eight states and some territories were being visited by the Commission's fish messengers.

The following year, 1881, is noteworthy for several reasons. In July James A. Garfield, the one-time congressman from Ohio who sponsored the original bill creating the Commission and now the twentieth president of the United States, was shot. He would succumb to his wounds within a few months. In that same month Booker T. Washington opened his Tuskegee Institute. A few months later the infamous gunfight at the O. K. Corral pitting the Earp brothers and Doc Holiday against the so-called Cowboys led by the Clanton brothers took place in Tombstone, Arizona. Though not recorded in history with nearly as much fanfare, 1881 also is the year when the Commission took possession of its first fish car.

The distribution of carp took top honors among the "noteworthy features" Baird listed in his annual report for 1881. A close second was "the construction of a [railroad] car suitable for distributing fish of all kinds *and an entire change in the methods of fish transportation*" (emphasis added).[3] Obtaining its own fish car marked a major inflection point in the Commission's management of the nation's fish stock. The fish car (and the ensuing fleet of cars) not only allowed for a more rapid delivery of fish, it also altered the way in which fish could be stocked. Fish cans and messengers traveling on the many railroads would continue to be sent out to fill orders, but the fish car gave the Commission the ability to transport and stock many times the number of fish handled by a fish messenger shuttling cans of fish between trains. From the perspective of fish culturists, the use of a fish car also greatly improved the odds of the stocked fish's survival. Later it would be decided that the fish car also could be used to ship larger fish than fry, increasing the odds of survival post-deposit even more. From a public relations vantage point, such an improvement translated into more fish to be harvested by local fishermen, whether for food or for sport.

The work of the Commission became more obvious to the public. Fish Car No. 1 was inscribed with "Philadelphia" on its letter board—the area above the windows—because it came from the Pennsylvania Railroad Company in Philadelphia. It also has been suggested that with this labeling stationmasters and agents of other railroads might give this car different consideration if they believed the car was owned by the Pennsylvania Railroad. More importantly from a public relations standpoint, the very obvious "United States Fish Commission" also appeared along the sides of the car. Think of it: perhaps you had read of the Commission's important work, and now you might catch a glimpse of one of its cars rolling along to another deposit site.

Similar to the California aquarium car, the Commission's No. 1, as it was referred to, had dedicated areas to hold ice; had areas for large tanks of water; had sleeping and living areas for the messengers (a maximum of five); and used some of the more recent innovations in rail travel, including Westinghouse air-brakes, a Miller platform, and six-wheel trucks (the structure beneath the car to which the wheels are attached). These innovations made the car capable of being attached to high-speed passenger trains instead of freight trains that often took much longer to get from point A to B. More importantly, it could hold enough cans and tanks to carry one to two *million* fish per trip.

Car No. 1 had a somewhat inauspicious initial outing, though not nearly as dramatic as the California aquarium car. After arriving in Washington DC on 7 May, the car's first trip was to deliver shad to Atlanta in early June. Unfortunately, soon after departing Washington, it encountered difficulties at a stop in Lynchburg, Virginia. The issues required some time to be resolved, so the shad were dumped into the nearby James River. Once repaired the car headed back to Washington. Its second "inaugural" trip occurred in mid-June. The car was loaded with over a million shad fry from the hatchery at Havre de Grace in Maryland. Its destination was Maine, where the fish were successfully deposited into the Kennebec and Mattawamkeag Rivers.

Even though No. 1 had made only one trip by mid-1881, the fee charged to the Commission for moving the car already had been established. In May President Hinckley of the Philadelphia, Wilmington & Baltimore Railroad—an early and staunch supporter of the Commission's activities—offered to move the Commission's fish car, its contents, and the attending messengers at a rate of 20 cents per mile. It wasn't free, but it was significantly lower than the going market rate. At the time the Pennsylvania Company was charging about $180 to haul a private car from Pittsburg to Philadelphia, a distance of about 350 miles. That translates to a rate of slightly over 50 cents per mile. Hinckley's charging the Commission 20 cents per mile was a welcome and generous gesture, one that Baird recognized in his *Report for 1885*.[4]

Hinckley's first move helped establish the going rate. Other major carriers fell in line—no doubt Baird negotiated these rates using Hinckley's as leverage—including some of the largest railroad companies in the country. These included the Baltimore & Ohio; the Chicago, Burlington & Quincy; and the Illinois Central. The Union Pacific and Central Pacific Railroads went one better and agreed to move No. 1 from Council Bluffs, Iowa, to San Francisco for $270 dollars, much less than 20 cents per mile. What makes the latter's agreement intriguing is that it faced no competition for this route. The other railroads, to some degree, were competing against each other for traffic and revenue. These agreements with some of the country's major carriers allowed the Commission to move its fish car (soon cars) on these railroads (and their branches), reaching a larger swath of the country more effectively than ever before.

Having taken the fish car and its crews through a couple of trial runs, and with rates of travel established, the real test for what the fish car could offer came in early 1882. The Commission decided to use No. 1 to deliver carp fry to applicants spread across the states of Arkansas, Louisiana, Missouri, and Texas. Applicants had received carp shipments before, but by using messengers. Now hundreds of applicants would get their fish *in one coordinated trip*.

A successful trip would generate many positive benefits. It would certify the ability of the Commission to readily meet the public's demand for fish, whether it was carp or trout or any other species. The Commission's reputation could only rise once it demonstrated its ability to effectively deliver fish, this time carp, on a timely basis. Being able to deliver the fish almost anywhere they were requested also would strengthen the Commission's status with the public and every local politician and official. Because the Commission required applicants to file their orders through their congressman's office, the more orders filed and successfully filled would bolster the Commission's standing with congressmen across the country. Aside from the benefits accruing to the Commission, the trip—if successful—also would raise the public image of the railroads who were assisting the Commission in this public-serving enterprise. Announcements of the fish car's arrival did, after all, include the railroad on which it was traveling.

By 1882 applications for carp were increasing rapidly: some 950 requests were filed from individuals in the state of Texas alone. Marshall McDonald coordinated the project.[5] The fish car would leave from Washington DC with a maximum load of fish, and, as the number of applicants' orders were filled along the route, it would be restocked from milk cans of fry with fish messengers sent ahead to prespecified locations by express delivery. The logistics required to achieve this delivery meant that McDonald had to determine the most efficient route for the fish car to take, requiring the fish car to travel to central points from which nearby applicants could pick up their fish, or from which the fish would be sent on by express. A fish agent stationed at these nodes would, in theory, ensure a correct distribution. All of this required McDonald to coordinate the movements of the fish car and the fish messengers to meet applicants at predetermined stops. Of course, it also meant coordinating with the railroads where and when the fish car would be arriving, departing, et cetera. This was no small feat since over the course of the route the fish car would by necessity be transferred between a number of railroad companies.

Details of the route were drawn up and notifications sent to the lucky applicants about where and when they could collect their carp. The plan was released to the public, ensuring the maximum public relations benefit. Once loaded, No. 1 left Washington DC on a cold 3 January and headed west on the Pennsylvania Railroad. After a few days it arrived at its first major stop, St. Louis. In St. Louis distributions of carp were made to applicants from Missouri and Iowa. The car then was transferred to the Iron Mountain Line and sent to Texarkana, Arkansas, where orders from applicants from Arkansas were filled. The car then made its way to Texas, stopping in Sherman, Dallas, Austin, San Antonio, Laredo, and Houston. It also made a stop in Shreveport, Louisiana, to fill orders from individuals in the vicinity. The empty car headed to St. Louis and from there back to Washington.

The trip was an outstanding success. Although there were some minor glitches along the way, the Commission had demonstrated to the public what it could accomplish with the fish car. The Commission demonstrated the practical efficiencies of using a fish car relative to the heretofore standard approach of using fish messengers to deliver comparatively small batches of fish. The trip also revealed how reliant the Commission was on the private railroad companies that shunted the fish car here and there. "The satisfactory issue of our work," wrote McDonald, "is largely due to the liberal facilities accorded us by the various lines of railroad traversed. Anything in the way of supplies or service was unfailingly rendered.... From Saint Louis westward until our return to that point, free transportation... was granted on all lines of railroad traversed by us."[6] It should be noted that for this venture the various railroads involved decided that the twenty-cent-per-mile charge was unnecessary.

The trip was quite a public relations triumph. All associated with it—the Commission, the railroads, and local dignitaries—claimed a degree of responsibility for its success. The public expressed interest in the fish car's abilities, and the arrival of the Commission's fish car with its cargo became a newsworthy event. An article in the *Dallas Herald* declared that the public should "appreciate the true value of this great work, and the incalculable benefit necessarily derived therefrom."[7] Though not quite as effusive, articles in the San Antonio *Evening Light* and the Austin *Democratic Statesman*, among others, praised the Commission for putting its fish car to such good use. Such complimentary articles of the Commission's carp experiment even found their way into major metropolitan newspapers. The *New York Times* in an 1880 article alerted the public to arrival of the Commission's fish car and one thousand carp being sent from Commissioner Baird to the New York State Fish Commission for distribution in local ponds. Four years later, almost to the day, the *Times* reported that another Commission fish car was due to arrive with another shipment of carp, noting "the success of the carp culture in this country has been very remarkable."[8]

An article that appeared in the *Opelousas Courier* is worth singling out. The *Courier* provided the Parish of St. Landry, Louisiana, with all the important news of the day. Why this article? I chose it because it highlights the symbiotic relation between the Commission and politics. The article per usual covers the arrival of the Commission's fish car and its delivery of carp to some lucky local applicants. The article also reveals why the Parish of St. Landry came to be blessed by such an event: the Honorable E. W. Robertson, the local congressman, "seems to be considerate in everything which may benefit his constituents," made it happen.[9]

Indeed, a detailed analysis of the Commission's role in introducing carp in the state of Texas, and the role played by the fish car, reveals that the Parish of St. Landry example can be multiplied manyfold. Politicians at all levels of Texas government—local, county, state, and federal—took their share of credit for the arrival of the Commission's fish car.[10] Given the outsized popularity of the carp program, my guess is that laudatory comments in the press about their political figures appeared in newspapers wherever carp were planted. And I am sure that Commissioner Baird was quick to remind those in Congress of the Commission's popularity when debates over congressional funding arose.

As you already know, the public's torrid love affair with the carp quickly cooled. Though the carp experiment was eventually halted, there still were shad, trout, a numerous other species of fish to be moved. As the Commission expanded its range of operations, the fish car would prove to be a phenomenal success.

The Fleet Grows

The Commission's success of moving carp thousands of miles "forcibly illustrates the improvement in methods of transportation introduced since 1881. By the methods in vogue prior to that date it would have required ten messenger shipments to have accomplished the same work, and would have involved tenfold the cost."[11] Marshall McDonald wrote those words in his review of fish distribution activities in 1883. By that time the fish car already had become a recognized fixture in Commission undertakings. It was so successful that Congress authorized the funds for a second fish car in early 1882.

The Baltimore & Ohio Railroad Company began building the newest fish car in March of 1882 and delivered it to the Commission at the end of May. The specifications of the second fish car came from Frank Eastman, an engineer officer of the Commission. The new fish car was different enough that the Commission published an article by Eastman in its *Report of the Commissioner for 1882* detailing the newest model. Fish Car No. 2 had "Baltimore and Ohio Railroad Company" displayed in its letter board as a gesture to the company who built the car. "United States Fish Commission" was displayed on the car's side and over the rear doorway to let the public know who really owned it.[12]

No. 2 was slightly longer and slightly wider than its predecessor. Like No. 1 it was about fourteen feet high. More importantly, the new car was designed to carry a heavier load (ten tons) than the first car and still travel at passenger train speed. This required its springs to be much stronger than those used for No. 1. To ensure that it could haul heavy loads at passenger train speeds, the new car also was fitted with Westinghouse brakes, a Miller platform, and the newest of couplers.

The novelty of this car was that the cans for holding fish were connected by a system of pipes and hoses to a pump and a blower located in a room at one end of the car. This innovation circulated water and air, respectively, through the cans and thus relieved the messengers from this tedious can-by-can chore. The pump and blower ran using a system of pulleys and belts connected to the truck of the car. There were also iceboxes for storage.

No. 2 provided greater comfort for the fish messengers. Chairs were suspended from the ceiling and, when needed, could be placed on top of the compartments holding the fish cans. Sleeping cots folded down from the ceiling. At the same end as the room with the blower and pump was a pantry and small kitchen, complete with a sink. There was also a washroom and a closet for a heater (if needed). For all intents Fish Car No. 2 amounted to a self-contained, rolling hatchery.

To make the car more adaptable to traveling across different railroads, additional expenses were incurred in the purchase of trucks that allowed the car to be switched from the 4 feet x 8 ½ foot gauge standard on Northern railroads to the more common 5-foot gauge used in the South. As discussed in chapter 1, this discontinuity in rail gauge required stopping to change out the trucks, increasing the costs of moving the fish cars when they traveled between Northern and Southern states. Trucks of the two sizes were stored at various stations in the South, where the switching of the fish car from one system to the other could be done efficiently. With all these and other special items, the total cost for No. 2 amounted to about $8,000, or close to $235,000 today.

No. 2's first trip took place in early November 1882, when it set out to deliver carp to various locales. It also was used to deliver shipments of shad during the year. The Commission made extensive use of its two fish cars in 1882. Fish Car No. 1 racked up over 30,300 miles delivering mostly shad and carp. Fish Car No. 2 logged over 16,300 miles, which actually is quite amazing, since it did not enter into service until late in the year.

The use of fish cars required the Commission to modify its now perfunctory annual recognition of the railroad companies providing assistance. Baird's 1882 commissioner's report makes the usual nod to those railroads that had "accorded the facilities for carrying fish in baggage cars and for stopping trains at bridges so as to deposit young fish." The list for 1882 includes ninety-seven different railroad companies and their subsidiary lines that provided assistance to the Commission. Sometimes it amounted to carrying fish messengers hundreds of miles, other times just a few. Even though the level of assistance varied, all were recognized. Baird's entry also includes for the first time a "LIST OF RAILROADS THAT MOVED CARS, AND MESSENGERS TO THE NUMBER OF FIVE ACCOMPANYING, AT THE RATE OF TWENTY CENTS A MILE DURING THE YEAR 1882."[13] Twenty-seven railroads made it to this list. What is interesting is that the two lists overlap: except for four railroads, twenty-three also are called out for ferrying the messengers and their cans for free or at reduced fares.

The practicality of the fish car would be demonstrated time and time again over the next two decades. The Commission, not surprisingly, asked for and was able to secure congressional funding to enlarge its fleet of fish cars. Fish Car No. 3 arrived in 1884 and featured a significant innovation by Frank Clark, superintendent of the Commission's hatchery in Northville, Michigan. Clark's design change made it possible to load the car with fertilized eggs at one of the Commission's hatcheries, hatch them en route, and upon reaching the destination have viable fry to deposit. An article in the *Chicago Tribune* praised Clark's enhancement, proclaiming that "nothing of the kind has ever been undertaken," and that the new fish car "will result in a great saving of the time now lost in returning to the hatcheries for fresh supplies."[14] Most important, the innovation actually worked. In one of its most significant trips in 1886, No. 3 took on six hundred thousand shad eggs at the Battery Station in Maryland destined for Portland, Oregon. The eggs were hatched along the way, and, by the time the car reached its destination, 2,800 miles later, the resulting fry were deposited into the Columbia and Willamette rivers.

The next addition to the fleet was a new baggage car bought from the Harlan & Hollingsworth Company in Wilmington, Delaware, in March 1893. No. 4 was fitted with the newest car couplers to abide by the latest railway regulations, equipped with four cedar water tanks, two of which were four feet in diameter and two that were eight feet in diameter. All were about two feet deep and bolted to the floor of the car. This car also had a device to better aerate the tanks. There were included the now-common sleeping berths and cooking arrangements. No. 4 could transport 150 ten-gallon cans containing up to 15,000 larger, three-inch fish (fingerlings).

A unique feature of No. 4 was its weight. Tipping the scales at nearly 43,000 pounds made it one of the heaviest railroad cars in the country. Even so it still could travel at high speed with passenger trains which helped reduce delivery times. Its weight sometimes created problems, however. The weight of No. 4 was even blamed for an accident in 1911 near Bridgeport, Connecticut, in which thirteen people were killed and forty-eight injured. The popular story is that No. 4 caused the Federal Limited, to which it was attached, to be delayed. When the engineer sped up to make up for lost time, the extra weight of the fish car allegedly caused the accident. Despite the damage the fish car survived the accident, perhaps leading to the blame landing on it. After an investigation the fish car (and the Commission) was absolved from any responsibility.[15]

The Commission's fleet of fish cars increased to five by 1900. Aside from their proven usefulness, another explanation for the growth in the fleet was the expanded system of federal fish hatcheries. In 1880 there were ten Commission hatcheries, or "stations," as they were called. (I am ignoring the Commission's marine operations at Woods Hole and other locations.) A decade later there were twenty-two. With more and more hatcheries, the fleet of cars simply had more product to move. From these hatcheries found across the country (interestingly, except for one in Virginia, there were none in the South) a total of over three hundred million fish eggs, fry, and adults were being distributed *annually* by the Commission. With the majority of its cargo being fry and adult fish, the Commission's reliance on the fish car—and on the railroads that moved them—was even greater than before.

Some Numbers

Table 1 provides some perspective on the Commission's fish delivery operations through the end of the 1800s. Several relevant bits of data are presented for select years: the number of fish cars, total miles traveled by these cars, and the miles traveled by fish messengers. As important as the fish cars became, notice that they did not totally replace the fish messenger and his cans as a method of delivering fish. To appreciate just what the railroad companies brought into this endeavor, where available I report on the miles provided by the railroads at no charge for fish cars and for fish messengers. Finally, in the far-right column is an estimate of just how many fish eggs, fry, fingerlings, and adult fish the Commission—and the railroads—was shipping. It is worth noting that in the later years the vast majority of the shipments were fish, not eggs.

Table 2. Data on use of fish cars, messengers, and scope of operations

		Total miles by		Total free miles for		
Year	Cars	Cars	Messengers	Cars	Messengers	Total Distribution*
1855	3	74,805	NA	26,212	NA	173,666,083
1890	3	75,312	50,009	41,339	NA	306,370,548
1895	4	93,377	75,384	65,189	16,389	619,825,852
1900	5	101,796	157,297	42,746	40,239	1,164,336,754

*Includes eggs, fry, fingerlings, and adult fish

This table reports data on the increase in the use of fish cars by the Fish Commission for select years between 1885 and 1900. It also shows the volume of fish distributed by the Commission.

Sources for Mileage data: *Report of the Commissioner for 1885*, CIX–CX; *Report of the Commissioner for 1889–1891*, 60; *Report of the Commissioner for 1895*, 51; *Report of the Commissioner for 1900*, 16.

Sources of Total distribution data: *Report of the Commissioner for 1885*, CV; *Report of the Commissioner for 1889–1891*, 13; *Report of the Commissioner for 1895*, 72; *Report of the Commissioner for 1900*, 15.

The Commission clearly used its fish cars extensively. This is illustrated by the fact that in the years covered the miles traveled per fish car averaged over 20,000 miles per year, about one trip around the Earth. The fact that the fish cars were in use generally from April through November makes the miles traveled even more impressive. The column reporting total miles traveled by messenger shows a notable increase over time. In just the last decade of the century, fish messenger miles more than tripled, from about 50,000 to nearly 160,000 miles per year. Fish cars increased the ability of the Commission to deliver various types of fish and of larger size to farther places. Fish messengers remained the backbone of its delivery service.

It is likely that the miles listed actually understate the distances racked up by the fish cars. This is because the Commission sometimes loaned their fish cars to states that did not have their own, and those miles are probably not included in the official tally. For example, the Illinois State Fish Commission notes in its *Report* for 1888 that in the spring the Mississippi River had overflowed its banks, leaving many fish in the ponds and sloughs that formed once the flood waters receded. The standard response to such an event was to use small boats to gather the marooned fish and use them as a supply of native species (e.g., bass and crappie) for distribution across the state—in effect restocking without recourse to artificial propagation.

Because the flooding was so severe and the numbers of stranded fish so large, the customary approach was not possible. The Illinois commission thus contacted Baird to ask if he would lend them one of the Commission's fish cars to carry out their redistribution of native fish. Baird agreed, putting not one but two of the Commission's fish cars and crews at the disposal of the Illinois commission, which used the fish cars from mid-July through early October. This incident also provides further evidence of the railroad's participation. "We never asked a favor [from a railroad] that was not granted," says the Illinois commission's report, nor "never desired a particular train that we were not given it, and water when needed, with every attention possible cheerfully accorded us."[16] This note of thanks was followed by a list of the two dozen or so railroads that aided the Illinois commission in moving the Commission's fish cars around the state.

The commissioner's annual thanking of the railroads who helped the cause changed during the 1880s. For the most part it became limited to acknowledging only those companies that moved its fish cars. The 1885 list included those who charged a rate of twenty-three cents per mile. Why this was an important distinction is unclear. According to the text, only the Union Pacific Railroad charged more, which was a change from earlier years. Only those roads that provided their service for free made the list in later years. Looking at the "Total free miles" columns in table 1, note that free miles are less than the total. This means that railroad companies were providing a service, but more of them charged some fee to do so. The price they paid was being excluded from the commissioner's annual list. Turning back to the table, the data also indicate that as time wore on, the percentage of total free miles traveled declined. For fish cars 1890 was the high point, with slightly more than half the total miles being provided at no cost. By 1900 that percentage fell to about 40 percent. Free miles for fish messengers amounted to about one-quarter of the total.

The total miles traveled by fish cars and messengers doubled in the last decade of the century. What also stands out is the fact that by the turn of the century the Commission was distributing over one *billion* fish in various stages of development (eggs, fry, fingerlings, and adults) to hundreds of locations across the country. And these numbers are for just one year! This monumental task represents a near-sevenfold increase from only fifteen years earlier, when "only" a little over 173,000,000 eggs and fish were being distributed. These figures indicate just how much the Commission expanded its operation. By the start of the twentieth century, its objective had shifted: yes, it was there to increase the supply of food fish to the nation's citizenry. But, more than that, it was now engaged in a full-blown attempt to manage the country's fishery. How that developed and its consequences are topics for another book.

As the new century unfolded, another technology would supplant the fish car and the fish messenger as the most efficient method to transport fish. During the 1920s the truck emerged as the principal means for transporting fish. A truck suitably equipped with water tanks could carry nearly as many fish as a fish car. More important, these special trucks cost less to operate and provided more convenient service to clients as the nation's network of roadways expanded. Applicants no longer received their shipment of fish at some train station: the fish truck could deliver their order directly to them. Even though the Commission continued to expand its fleet of fish cars—the last car, No. 10, entered service in 1929—the reign of the fish car and the fish messenger was coming to a close. After a long and highly productive run, the fish car era officially ended in 1947, when the last fish car, No. 10, was retired from service.

Fish Cars for State Commissions

The story of the fish car would not be complete without recognizing the role played by state fish commissions. Most had the same goals as the Commission. Often in cooperation with the Commission, many states jumped at the chance to introduce new species of fish into their local waters. (Recall the list of species introduced to California found in chapter 4.) And, like the Commission, states relied heavily on railroads to move their own fish messengers and cans of fish around their state. Some even acquired their own fish car.

From the records I could find, state commissions argued that fish cars, as demonstrated by the Commission's work, offered a technically better method of moving hundreds of thousands of fish from hatchery to stream. The Nebraska fish commissioners argued their case this way:

> The planting of this number of young fish [over seventeen million] in the ponds, lakes, and rivers in all parts of the State is a work of no inconsiderable magnitude and danger, especially when surrounded, as it is, with manifold difficulties and trials in their transport railways in shipping cans for long distances (in

many cases several hundred miles), often requiring two or three days en route, exposed to the casualties of storms, and delays and sometimes losses of the whole consignment.[17]

A fish car would alleviate these difficulties. And having one would allow the commissions to distribute larger, more viable fish like the Commission. The Nebraska commission noted that the technology of using milk cans to deposit fish constrained them to plant fry that "are too young and feeble to take care of themselves . . . [which] are often nearly and sometimes entirely annihilated" soon after they are released.[18] This rationalization is common to most other commissions' appeals to their state legislatures to fund the purchase a fish car.

State commissioners also used the political angle to argue their case. The Nebraska commission noted in its report that "if the resultant effect [the increased efficiency in depositing fish across the state] could be to *increase the benefits to the people* many fold over present methods, as it is confidently claimed, it will be money well expended" (emphasis added).[19] When it announced that it had (finally) acquired its own fish car, the Iowa commission echoed their neighbor with the observation that "such a car [would be] very essential to *the work of the commission in supply food fish for the people*" (emphasis added).[20] Because a fundamental objective of any fish commission was to increase the supply of food fish, why not use this lever to wrest funds from state legislators? If state fish commissions were not as successful as state residents would like, the lack of fish car could be pointed to as a key reason.

The states that acquired a fish car (or two) before the turn of the century are listed in table 2, which includes the year in which the car was put into service and, where available, its name and cost. California was the first state to acquire a fish car, its "celebrated" though ill-fated aquarium car. Lost on its first outing, it was replaced in 1874. The rest of the states acquired their fish car after 1880.

Table 3. State Fish Cars

State	Year	Name	Cost
California	1873	NA	NA
	1874	NA	NA
Iowa	1896	*Hawkeye*	$1,600
Michigan	1888	NA	NA

Missouri	1882	NA	$842
	1885	Walton	$2,000
Nebraska	1889	Antelope	$2,000
New York	1891	Adirondack	$3,800
Ohio	1891	Buckeye	NA
Pennsylvania	1892	Susquehanna	$5,000
Wisconsin	1893	Badger	4,500

This table shows the states and the year in which they acquired their first fish car. The time covered is 1873 to the early 1890s. The name of the car and its cost, if available, also is reported.

Source: Various state commission reports.

You might think that New York or Pennsylvania would have been the most likely after California to procure a fish car. After all, fish culture had an established history in those states, and they were states that established fish commissions early on. In fact, Missouri was the second state to have a fish car. This was big news. *Forest and Stream*, one of the most widely read purveyor of all things sporting, ran a story in the summer of 1881 announcing that "the Missouri Commission is about to fit up an aquarium car to transport fish."[21] It is interesting—and telling—that a national publication like *Forest and Stream* bothered to report this event. Sadly, Missouri's first fish car was as cursed as its California cousin. It had hardly seen any service when only a few days into the new year of 1883 it was destroyed by fire. It had the misfortune of being parked near a grain bin in St. Joseph, Missouri. When the grain bin caught fire, it spread to the idle fish car. Missouri's fish commission lobbied the legislature for a replacement, and, though it took some time, the *Walton* was put into service a couple of years later.

Other states joined in the fish car fraternity. Michigan and Nebraska acquired their fish car by the end of the 1880s, with other states soon to follow.[22] From the available descriptions, all were built along the same specifications as the Commission's first cars. Indeed, looking at renderings or photos of the state cars, it is difficult to tell them apart. Like the Commission, to ensure that the public knew what it was, the name of the state's fish commission was emblazoned on the side, along with the car's name, if it had one. Though similar in dimensions, apparently there were enough differences to warrant disparate price tags. Of those listed, Pennsylvania's *Susquehanna* was the most expensive, followed by Wisconsin's *Badger*.

One thing I have not been able to locate is consistent data on the miles traveled by state fish cars. Even so there is enough anecdotal evidence (e.g., the passing remark by a fish commissioner of making dozens of trips with the fish car) to suggest that the miles racked up by state fish cars was not inconsequential. And, from the state reports I have read, in almost every instance the local railroads often moved the cars around the state for free. Like the Commission, state commissions' annual reports made sure to acknowledge those railroads that provided support.

It bears emphasizing just how much the state commissions relied on railroads for assistance. In one of its annual reports, the Illinois Fish Commission effusively thanked the railroads, who "without their aid given just as it was and at the time it was, we would have failed even though all other interests were favorable." They also recognized the importance of the state's popular press in alerting the public to the fish commission's activities. "By their ready cooperation and the generous use of their columns when needed," the Commission stated, "we have been able to reach the people with our announcements of distributions, etc. . . . [and] have found them [the press] as ready always to bestow praise as censure, if we deserved it."[23]

The press also bestowed recognition on the local railroads for their support of the work done by the fish commissions. To use one example, the *Waterloo* [Iowa] *Courier* reported that "the Illinois Central railroad company has been doing all in its power to aid in this distribution of fish in the lakes and streams along its route," adding that the company was moving the Iowa Commission's fish car, the *Hawkeye*, for free.[24] Sometimes the press even provided an outlet to bring public pressure on recalcitrant railroads. A wonderful example of using the press to influence public opinion is the reprimand of the Erie Railroad Company by Henry C. Ford, the president of the Pennsylvania Fish Commission. In the *Milford* [Pennsylvania] *Dispatch*, Ford is quoted as saying, "The Erie road . . . is unwilling to give passage to our [fish] car unless at a charge of 20 cents per mile. This charge, with our limited means . . . will be prohibitory. As the stocking destined solely for the benefit of the country adjacent to the railroad, and as every railroad in our own state except the Erie has granted our request for free transportation, we think the refusal will be injurious to the Erie in checking the progress of the fishing interests along the line."[25] The

Erie's reluctance to provide free service to Pennsylvania's fish commission was, according to Ford, an affront to the public interest.

Ford's note contains a harbinger of an issue that I will deal with in a later chapter. Ford argues that the Erie's refusal to provide free service to the commission would slow the growth of sport fishing along its line. This insight reflects the fact that the cooperative relation between railroads and fish commissions often meant that railroads experienced increased passenger traffic when the public realized that the track-side streams were abundant with fish. If the Erie did not help the commission plant fish in the streams along the Erie's tracks, fishing in those streams would not become popular, and the Erie would not realize the revenues from increased passenger traffic of those taking to the outdoors.

A Final Note

If one considers the entire period over which the Commission was transporting fish by rail, the numbers of fish moved and miles traveled are simply astounding. According to one estimate, "between 1872 and 1940 [the effective end of the fish car era] more than 200 billion fish and other aquatic species were distributed, largely by fish cars traveling 2,029,416 miles and detached messengers traveled an additional 8,104,799 miles with their fish cargo."[26] Whether by fish car or by milk cans accompanied by fish messengers, the sheer volume of fish shipped around the country attests to the Commission's commitment to actively manage the nation's fish stock—and also to the vital role that railroads played in achieving its policy goals. And let's not forget that the railroads also were actively assisting state fish commission as they carried out their own fish management policies.

The marriage of two technologies—rail transportation and fish culture—led to profound results. The fish car led to a marked improvement in the Commission's ability to stock fish. The fish car allowed federal and state fish commissions to significantly expand their enterprise, not only in terms of the sheer number of fish deposited but also in terms of geographical reach. The Great Experiment expanded to include dozens of different species being stocked in nearly every corner of the country. And it was all due to the railroad's infrastructure and the companies' willingness to lend their assistance. As you will see, the railroads' participation in this experiment and their marketing campaigns to increase ridership would significantly affect the arc of sport fishing in America.

RAILROADS, THE LANDSCAPE, AND SPORT FISHING IN AMERICA

Railroads were instrumental in allowing the federal and state fish commissions to develop and expand their fish-stocking activities during the latter half of the 1800s. With improvements in transportation technology—faster locomotives, heavier cars, and the specialty fish car—railroads increased the geographical reach of the commissions' fish-planting experiments and allowed them to significantly increase the number of fish planted. For my purpose an important outcome of this railroad-fish commission partnership was that it made available to anglers at all levels of skill the kinds of fish that they had only read about. This was important in building the foundation for continued growth in the emerging conservation movement. It also would engender sport fishing in America.

This and the next chapter will examine other ways by which railroads can lay claim to advancing this recreational pastime and nascent industry. This chapter specifically highlights ways in which railroads made nature and the activities if offers, like fishing, an appealing and popular form of entertainment. In doing so railroads adopted a proactive role in protecting and improving the landscape through which their lines ran. It made perfect sense: without fish to catch or beautiful surroundings to enjoy, there would be fewer customers. Before I delve into those topics, however, let's step back and take a broader view of how railroads encouraged a new and growing activity known as tourism.

Railroads and Tourism

A feature of railroads that early advocates pointed to was their ability to allow urban residents to escape to the countryside. The ability to get out of town would restore one's health from the damages—physical and mental—that many considered a byproduct of urban living. Given the nature of employment at the time, such trips to the country often were enjoyed only by those who could afford to take the time away from work. One student of the times noted that "the poorest members of society, of course, lacked both the time and the resources for such ventures. Day laborers, struggling journeymen, and domestic servants could not have taken off on 'tours' to Newport of 'jaunts' to the springs. Neither were farmers able to leave their land for extended summer excursions to summer resorts."[1] During the first half of the 1800s, even those individuals employed in what today are considered middle-income professions, such as law, education, and business, rarely enjoyed what we would call a vacation.

That all changed following the Civil War. As the economy healed from its scars, especially the economy of the North, incomes rose and with them the emergence of a socioeconomic middle class. Rising incomes and changes in the labor market led to an increase in leisure time for many households. Though railroads had already been providing access to various resorts for the wealthy prior to the war, rising incomes and improved rail access, including the newly opened lines to the West, were catalysts leading to a boom in tourism, or travel for travel's sake.[2]

This rise in tourism came about because of several changes. One change was railroads laying additional track to existing and new resorts. For example, White Sulphur Springs in West Virginia was for many years a popular escape for wealthy Washingtonians. Prior to the Civil War, reaching the resort may have taken the better part of a week of uncomfortable coach travel. That changed when the Chesapeake & Ohio Railroad built a line that dropped passengers off near the resort's front gate. This reduced travel time for many to less than a day's ride by train.[3] With the significant drop in travel costs, more individuals could and did visit the resort. Similar stories can be told for other resorts, including those in the White Mountains of New Hampshire; the Blue Ridge Springs in Virginia; the Greenway Hotel on Green Lake in Wisconsin; and Hot Springs, Arkansas, to name a few. Railroads increased the demand for such vacations by providing easier access. In doing so railroads had found a new source of revenue for which to compete: the tourist.

Some railroads even created tourist destinations of their own to capture more of the tourist-related revenue. Leaving the details for later, one example is the Hotel Del Monte, located on the Monterey Peninsula in California. The Pacific Improvement Company, part of the Southern Pacific Railroad, built and managed the hotel. Instead of luring customers to one hotel, the Atchison, Topeka & Santa Fe Railroad promoted the town of Santa Fe, New Mexico, as a tourist destination. Those railroads that could promoted travel to destinations like Colorado and the newly minted national parks.

As the century wore on, the growing number of jobs with steady salaries and employment expanded the middle class. Taking vacation time became an increasingly common experience. The confluence of comparatively affordable travel costs (in terms of ticket prices and reduced travel time) and increasing household incomes increased the number of tourists. To this segment of the traveling public, railroads directly advertised a variety of getaways. Billboards and posters of the day provided enticing illustrations of the destinations to which passengers could travel, even if for only a day or two. Tourist guidebooks published by railroads and private companies provided not only information on where and at what time their trains ran, but also about scenic sights, hotels, restaurants, and even guide services for summer escapes along their lines. Competition among railroads for the tourist dollar became fierce.

I think it is safe to say that without railroads tourism in the United States would not have developed and flourished as it did in the latter half of the 1800s. An unintended consequence of railroad's promoting tourism and travel was a deep sense of national pride. "Railroad travel in the West," one author observes, created "a pattern in which the act of tourism affirmed American society."[4] More and more wealthy tourists who once eschewed domestic travel for the European "tour" were now heading to the American West. The mountains of the West became our Alps. Other areas were similarly compared to European destinations. *Harper's New Monthly Magazine* claimed that "many thousands flock in steadily increasing numbers to the Northwestern [meaning Wisconsin] lakes and forests [where] one is strongly reminded ... of the sunny valleys and noble forests of Germany."[5] Foreign tourists, especially the British, became increasingly drawn to the natural wonders of the western United States. "No place on the earth's surface," notes one historian, "seemed to offer more all-round attractions to the British than the American West."[6] Colorado, it seems, was an especially popular destination for the British tourist.

Just how much did the railroad's push to increase tourism affect their passenger service? The data are not as complete as one might wish, but there are enough to get a feel for the changes that took place. Consider that in 1876 industry revenue from the passenger side of the business amounted to a little over $136 million. This figure doubled by 1890. And the average tourist saw travel costs decline in real terms: the nominal price for a ticket had been kept down because of stiff competition. A related piece of evidence of the rise of rail travel is the number of passengers. During the 1880s alone, the number of railroad passengers almost doubled, rising from 289 million in 1882 to 520 million only eight years later. There were more passengers, and they were able to travel longer distances: passenger miles jumped by two-thirds during the 1880s, from almost 7,700 million miles in 1882 to a little over 12,500 million miles by 1890.[7] During the last quarter of the nineteenth century, the increased miles of railroad track crisscrossing the country meant that train travel had become the only reasonable mode of transportation if one traveled any significant distance. "Before Henry Ford popularized the horseless carriage," notes railroad historian John Stilgoe, "any overland trip more than twenty miles involved a train. Period."[8]

The popularity and growing ease of train travel translated into increasing numbers of anglers visiting the nation's rivers, streams, and lakes. Francis Austin notes in his iconic book *Catskill Rivers: Birthplace of American Fly Fishing* that "all the money in the world would not have mattered [to making fishing popular] had there been no practical way to reach the rivers . . . and introducing masses to the sport was played by the railroads."[9] As you will see, the railroads' influence went much deeper than simply getting anglers to this or that fishing spot.

Advertising

One explanation for the increase in passenger traffic is simply that railroads began to aggressively promote this side of their business.[10] In the early years a railroad's timetable often served as its main form of advertising. Timetables appeared in newspapers and presented the basic facts: when trains were scheduled to leave and arrive at various stations. Some of the early timetables might include an image. In one for the Rutland & Burlington Railroad a simple graphic of a locomotive pulling three cars behind it appears over the schedule. Some timetables were printed as portable cards. The Mount Washington Railway, for example, had on one side of the card its schedule of departures and arrivals, and on the other was a colored image of Mount Washington. As with any form of advertising, the idea was to inform the customer and differentiate your product from the competition.

As railroads expanded their reach, number, and importance, the simple timetable by necessity evolved into a more comprehensive document. Because an increasing number of railroads were forming connecting routes with others, timetables and schedules became quite complicated affairs. Not only was there simply more information to convey about departure and arrival times and routes, but businesses along the railroad's route began to advertise in them. Some ads appeared for companies that were just consumer oriented. As important, the covers of schedules often illustrated what passengers might expect to see along the route or find at their destination. Nature often was the theme of the covers.

The push to encourage western travel led companies like the Atchison, Topeka & Santa Fe Railroad to become early users of advertisements in eastern and national magazines and newspapers. This advertising aimed to shift the wealthy eastern tourist's taste from traveling to Europe to exploring the West. As mentioned earlier, the pitch was that the U.S. West offered just as interesting and majestic scenery as anything Europe had to offer, and you could get there much cheaper and faster. By the early 1880s major railroad companies like the Union Pacific, the Chicago & Northwestern, and the Southern Pacific were running tourist-oriented advertisements in major magazines and newspapers.

As the West became even more accessible following the completion of the second and third transcontinental lines, railroad advertisements targeted individuals who hoped to realize a piece of the western promise. Railroads met the demand to move west and settle the new frontier by selling parcels off the land that had been granted to them by the government. That was the idea behind the land grant: railroads would sell their land grants to pay for construction, and the government did not care to whom they were sold. Some railroads thus offered "land seeker" tickets, which were sold at a deep discount compared with the normal passenger fare. The catch was that in order to have your land seeker ticket validated for the return leg, if one existed, the passenger would need to travel to the end of the railroad's line, or at least as far as where the railroad was trying to dispose of its land. Not surprisingly, a railroad land agent would be present at that depot and be more than willing to entertain inquiries about available land.

Over time railroad advertising began to focus less on times and dates. Railroad timetables and schedules could be acquired from ticket agents, as is true today: I can still visit the local train station and pick up an Amtrak schedule, though I am more likely to visit the company's web page and download one. As important as the schedule was and still is, railroads' marketing and promotions increasingly began to focus on the experience of rail travel and "the appeal of the place."

Throughout the history of railroads in America, one aspect of their advertising campaigns was promoting the "experience" of rail travel. And by experience I do not mean just the relative comfort or timeliness of the trip. Like the anticipation of early airline travelers a century later, the *experience* of riding the train often was as important as the destination itself. Railroads thus made sure to promote the scenery through which their tracks ran to attract passengers. As such railroad companies often treated the landscape viewed from their train car window as a valuable asset to their business, something like the modern notion of branding one's product. Some have suggested that rail travel created a disconnect between the passenger and the environment: the landscape and the people living in it whizzed by far too fast to enjoy or understand them. I think it is much more likely that railroads and their advertising served two purposes. One was to introduce individuals to destinations that they would never have seen were it not for a trip on the train. The other and more lasting purpose is that railroads helped establish an appreciation among the public for the environment that they would not have had if they had not taken the train.[11]

Railroads advertised the scenery along their routes using a variety of media: posters, magazine ads, timetables, guidebooks (of which I will have more to say later), and even playing cards. The archetypal poster, for example, represented an artistic rendition of reality. Depending on the company, the poster might depict rolling hills, forests, or nearby streams around and through which the train traveled. Some railroads went to great lengths to produce such advertising art, for that is what it was. Some even commissioned well-known artists to produce the images used in their posters. The painting *The Lackawanna Valley* (1845) was one such image commissioned by a railroad company and used to advertise its service through the area. Painted by the prominent landscape artist George Inness (1825–94), this is one of many examples in an influential body of American landscape art that incorporated trains and railroad infrastructure (e.g., bridges and viaducts).[12]

Using images of the scenery through which the lines ran exemplified one way that railroads suggested that their business and nature were compatible. The Baltimore & Ohio Railroad conducted one of the more outlandish—and effective—promotion along these lines. In June 1859 the railroad treated several dozen "artists and literati" to a four-day, round-trip excursion from Baltimore, Maryland, to Wheeling, West Virginia. A special train assembled to cater to the needs of this diverse group of passengers: one car for the photographers; a dining car, complete with sofas and a piano; two other cars with desks for the writers and artists; and a perfunctory smoking car that brought up the rear.

The trip was a public relations bonanza largely due to the fact that the B&O made sure that a correspondent for the popular national magazine *New Harper's Monthly Magazine* was invited. The *New Harper's* writer, in the flowery prose of the day, gushed that it marked "the first time in our history [that] the great embodiment of utilitarianism [the train] extended the hand to the votaries of the beautiful, claiming brotherhood and asking co-operation."[13] The thinly veiled moral of the article? Achieving the technological miracle of rail travel did not necessitate abandoning the aesthetics of an artistic life: technology, art, and nature could coexist.

The trip and the *New Harper's* article served two complementary purposes. First, it advertised the comfort in which one could travel and the interesting sights one could see and visit. Second, it uncovered to the broader public the beautiful landscape through which the B&O traveled and valued. Yes, building a rail line inflicts some environmental damage, but the scars soon heal over, and the glories of nature quickly return. As the *New Harper's* writer waxed poetic, the track was "sweeping and circling with the graceful sinuosities [sic] of the river" or "darting straight through a projecting spur; now under the cool shadow of a beetling cliff, then gayly emerging into sunshine and open fields."[14] This theme, not surprisingly, was carried on in the subsequent works produced by the artists and writers that made the trip. Such boosterism promoted travel on the B&O.

The idea that taking the train got you closer to nature is exemplified in a promotional scene that appeared on the cover of the 22 July 1893 edition of *Harper's Weekly*. It was a painting by the noted American artist Frederick Childe Hassam which showed passengers seated in the observation car—a car usually at the rear of the train with an open-air seating area—connected to a train bound for Chicago. The picture shows a mix of passengers, men, women, and children, enjoying what could be experienced only by taking the train: an expedient mode of travel, a diverse passenger list, and the added bonus of an unobstructed view of nature as it passes by. The not-so-subtle visual nudge is that if you want to travel in comfort and enjoy the sylvan beauty of nature at the same time, take the train. A more subtle nudge is that train travel is safe for everyone, including women and children. More on that in the next chapter.

Railroads tempted the urban dweller with an easy escape from the daily grind, the stress of city life: take the train and reconnect with nature. This argument, if you recall from chapter 1, was made for building a railroad system by its early advocates. And it is one that reappears in railroad marketing materials not only through the 1800s but even to this day. A popular guidebook from 1881—though it could have been from last year—declares that the train traveler can "gain new vigor by simple contact with nature, breathing the air . . . seeing the sights, and hearing the sounds of the country."[15] In an ad for the California Zephyr, the copy reads: "Experienced travelers say the California Zephyr is one of the most beautiful train trips in all of North America. As you climb through the heart of the Rockies, and further west through the snow-capped Sierra Nevadas, you may find it hard to disagree."[16] Then as now train travel has always boasted of the sights awaiting the traveler.

A more modern assessment of what benefits awaited the train passenger suggests that "the railroad offered release from modern anxieties. By delivering passengers quickly to the beauty of nature, the railroad provided an antidote to the enervating forces" of urban life.[17] The view of fishing historian Paul Schullery is that the train ride to upper Michigan or Colorado became "an important part of the fishing experience." So important, in fact, that articles about fishing trips appearing in the leading magazines and newspapers of the day were as likely to detail the train trip as they were the fishing itself.[18]

Preserving the Landscape

The idea that the scenery should be treated as an asset "brought leadership to the idea of preservation," acting in ways that demonstrated what to some seemed like a non sequitur: that "commerce and the landscape could coexist."[19] Railroads often took actions that protected the landscape through which their lines ran. (I will have more to say about this in the next chapter.) We know, for example, that the New York Central Railroad prohibited the placement of outdoor advertising signs along its tracks. When local entrepreneurs put them up, the railroad sent a crew out to tear them down. This railroad's reaction to outdoor billboards stirred public support for similar efforts. Arthur Reed Kimble opined in an 1898 article appearing in the widely read magazine *The*

Outlook that "if 'soulless corporations' are thus found willing to set the example of ridding their lines of travel of obnoxious advertisements ... what may not be hoped from individuals and communities when once the popular aesthetic conscience has been touched?"[20]

Railroads not only controlled what appeared along their tracks, but many also worked to preserve the landscape itself. Instead of maintaining clear-cut rights-of-way, they were allowed to return to their natural state. This provided what today we would call a "green" buffer between the tracks and the surrounding land. While many farmers were actively eradicating trees and native plants that impinged on their ability to maximize output from the land—today it appears as fence-to-fence farming—railroads were not. Railroad rights-of-way created their own environments that flourished with native wildflowers, bushes, and plants. There even was a "railroad gardening" movement in which railroads maintained their rights-of-way and created flower gardens around their stations "for the benefit of passengers and for the stationary observer gazing at the passing trains." After all, "good stations and grounds attracted passengers, and so made good business sense."[21] This behavior also created a positive externality for the public.

I will dig into the topic a bit deeper in the next chapter, but it is worthwhile to mention here that railroads often acted in the public good when it came to the environment. A history of the Southern Pacific Railroad describes numerous areas in which the company became a community leader in conservation practices. "In the area of water policy," writes Richard Orsi, "enlightened corporate self-interest led the Southern Pacific to identify with 'public welfare' and to work vigorously to solve water problems" in those regions in which they operated.[22] The Southern Pacific even worked closely with prominent leaders in the conservation movement, such as John Muir, supporting landmark decisions aimed at protecting the wilderness. "Think of it," urges Alfred Runte, "these were the railroads, America's most hardened corporations, talking about believing in natural beauty."[23]

Not losing sight of the fact that railroads were profit seeking enterprises for whom the goal was to attract passengers and enhance revenue—"enlightened corporate self-interest," Orsi calls it—the fact remains that the railroads' actions supported, directly and indirectly, the conservation and preservation of the landscape and, by extension, wildlife. By connecting passengers with the landscape, railroads also played an important role in the public's growing interest, especially among urbanites, in the outdoors and in the activities it offered, like fishing.

The Discovery of Place

The scenery along the train tracks was important. To grow the tourist side of the business, and to steal customers from the competition, railroads increasingly began to deliver passengers to popular destinations. During the decades preceding the Civil War, such destinations of note—many quaintly called "watering holes" by the press—were located mostly in the Northeast: Niagara Falls, the White Mountains, the Adirondacks, Saratoga Springs, and the Catskills.[24] These were the retreats of the wealthy—until railroads found their way into these once remote regions. With the tracks came tourists from a much broader swath of society. As passenger traffic to these established destinations increased, so did the number of places to visit. Many towns along a railroad's route to, say, Saratoga Springs began to advertise themselves as rural getaways, entrepreneurs offering houses for rent like the modern-day Vrbo or Airbnb. Some even offered meals and amenities like guide service to the local lakes and streams. Railroads were helping build the infrastructure of the nascent tourist industry, a subset of which would involve sport fishing.

Railroads democratized traveling for pleasure and sport. Many of the now accessible destinations offered the angler—both neophyte and seasoned—the chance to try his or her luck in waters that perhaps they had only read about in the past. Many of the old guard saw only the downside to this incursion into their once secluded haunts. The storied rivers of the Northeast were now accessible by more and more anglers and with this access arose a widening interest in fishing. Other locations, like the Pocono Mountains in Pennsylvania, also grew in popularity as a family retreat. Families from across the social spectrum could, after a relatively short train ride from some of the nation's largest metropolitan areas, enjoy a wide variety of outdoor activities, including fishing.

The Erie Railroad was the major provider of train service to the Pocono area. The Poconos were so popular that the lure of significant passenger revenues encouraged the Delaware, Lackawanna & Western Railroad, more commonly known as the Lackawanna, to enter the tourist trade. It is a good example of how railroads reacted to market opportunities, sometimes changing the focus of their business.

The Lackawanna started as a railroad that hauled coal from the Pennsylvania fields to New York: passenger service definitely was not a priority. But it had several stops located in Monroe County, Pennsylvania, home of the Delaware Water Gap. Known as the gateway to the Poconos, the Gap became (and still is) a popular summer retreat for visitors from New York City and Philadelphia. Once Lackawanna executives realized the potential revenue to be earned by serving those interested in visiting the Poconos, it began offering service to the area. By the mid-1860s, the Lackawanna had significantly reduced travel time to the Poconos, and within a few years it boasted of five trains leaving New York City daily for the Pennsylvania mountain resorts.[25] By the 1880s it, not the Erie, had become the major carrier of tourists to the region.

The Lackawanna made the scenic countrysides of the journey and what awaited the passenger major features in their advertising campaigns. They hyped the fact that the Poconos offered excellent fishing. With the decrease in travel time and expense, more and more individuals—men and women—escaped the city for the countryside to try this new pastime. With easier access via rail travel, the public's interest in sport fishing as a recreational activity, not merely a means to put food on the table, grew. With social standards relaxing from the constraints of the Victorian era, women were encouraged to take part in "healthy" outdoor activities. Fishing was among those activities considered to be a socially acceptable endeavor for a woman. (Many thought hunting too brutal for women, however.) By lowering the cost of reaching the Poconos and other fishing haunts, railroads were effectively lowering the barriers of entry into this new pastime. As Francis Austin noted earlier, railroads helped greatly to expand the pool of individuals who were taking up sport fishing.

The transcontinental railroads elevated the railroads' marketing of landscape and destination to a whole new level. The Southern Pacific Railroad was a pioneer in this area. As mentioned earlier, the Pacific Improvement Company, part of the Southern Pacific, built the Hotel Del Monte in Monterey, California. The Del Monte opened in 1880, offering lavish accommodations, service, and the most modern of amenities—for example, telephones. The Southern Pacific advertised the Del Monte as the "Queen of American Watering Places," and it worked. Wealthy Americans, including the likes of Andrew Carnegie and Joseph Pulitzer, who might have otherwise traveled to Europe, now went to the West Coast to visit the Del Monte. And, of course, the Southern Pacific, in their advertisements, made sure to mention the A-list guests. But the Southern Pacific did not only cater to the wealthy. For the less affluent traveler, the Pacific Improvement Company offered accommodations at its Pacific Grove. The Grove had more of a camp orientation, where families stayed in tents and cottages. Whether one was looking for a high-end or a more camp-oriented vacation, the Southern Pacific could deliver the tourist looking to escape to the West Coast.

To compete for the tourist dollar in the 1880s, the Atchison, Topeka & Santa Fe Railroad began to advertise Santa Fe, New Mexico, as an exotic destination.[26] Unlike the lavish Del Monte, the town of Santa Fe offered more accessible lodgings, thus appealing to a wider market. It also had the allure of the Wild West, including the presence of Native Americans. Promotional campaigns emphasized the region's unique landscape, totally unlike that found in the East or Midwest. The Santa Fe Railroad also marketed Santa Fe the town to those travelers seeking healthful environments, due to its arid climate.

Because reaching Santa Fe (or the West Coast if one chose to continue on) could entail a journey of several thousand miles, the Santa Fe Railroad offered amenities during the trip that one might expect at a posh resort. These included high-end accommodations and meal service expected by some of its clientele. But the railroad went further, adding touches of sophistication and pampering. When the westbound California Limited reached its namesake border, for example, every woman passenger received a small bouquet of roses, lilies, or violets. Male passengers were given an alligator skin wallet. On top of its pledge to provide first-class service, the company suggested in one of its pamphlets that passengers would travel with "persons you like to meet— successful men of affairs, authors, musicians, journalists, 'globe trotters', pretty and witty women and happy children."[27] Interestingly, the Santa Fe used almost the very same advertising pitch in some of its ads from the 1950s, though by then the women passengers were more than just "pretty and witty."

While some railroads built lavish hotels to attract wealthy customers, others built amusement parks and more modest campgrounds to appeal to day-trippers and campers looking for some time in the outdoors. Railroads were responsible for the development of a number of poplar escapes in Western Pennsylvania. The oldest amusement park in Pennsylvania, Idlewild Park, was built in 1878 by the Ligonier Valley Rail Road which was owned by the Mellon family. The Pittsburgh & Lake Erie Railroad established the more rustic Aliquippa Grove about twenty miles outside of Pittsburgh around 1880. There are others, and in each instance railroads were responsible for building retreats where families could find relief from urban life. These parks, "strategically plotted along transportation networks and among diverse regional landscapes," became popular escapes that "attracted populations of pleasure-goers."[28]

The Frisco Railroad built its own fishing lodge, named Forest Lodge, near the Gasconade River in Missouri.[29] Those headed for the Clubhouse, as it was known, would disembark from the Frisco at the specially built depot (Schlict Station) on the Frisco's tracks and from there walk or hire a wagon for the last mile to the Clubhouse. The lodge was near Schlict's Mill, a privately operated mill on a spring stream that fed the Gasconade.

The Clubhouse was a very popular retreat for several reasons. First, its proximity to cities in the area—St. Louis and Springfield—encouraged urban residents to visit for the day. Second, the Frisco Clubhouse offered accommodation that allowed extended stays and thus attracted families not only from Missouri cities but also from surrounding states. Third, the Frisco heavily promoted the lodge (and Schlict's Mill) because of the fishing opportunities it offered. The Gasconade River and the ponds around Schlict's Mill were popular fishing destinations in part because the Frisco worked directly with the Commission to stock rainbow trout. Well into the 1900s the Frisco encouraged residents in cities and towns across the region to visit the lodge, enjoy the countryside, and try their hand at catching that "exotic member of the finny tribe," as rainbow trout were often referred to in the popular press at the time.

Railroads promoted a variety of destinations—some relatively close by and others hundreds of miles away—to the adventurous traveler. Some railroads made it very clear about what they offered the interested traveler. The Grand Rapids & Indiana Railroad, for instance, confidently advertised itself as "The Fishing Line." Why? Because it could deposit the traveler to, as one of its ads reads, "famous summer resorts and lakes of northern Michigan," where the interested angler could enjoy some of the country's best fishing. It hyped the fact that by taking its trains—you could connect to it from many different lines across the eastern United States—you'd end up in the only place in America where you could catch the Michigan grayling. Railroads thus became as adept at transporting the wealthy to their classy hotels and grand resorts as they were at getting families and outdoorsmen and women to less fashionable camps and rentals, regardless of whether it was in the East or the Midwest or on the West Coast.

How do you respond to the following statement: "Yellowstone National Park exists because of a private company's pursuit of profit"? If you think that it is nonsense, you'd be wrong. Yellowstone probably would not have become the nation's first national park were it not for the early and enthusiastic intervention by the Northern Pacific Railway. "They [the railroad] favored the park," writes historian Paul Schullery, "because it would provide them with additional passengers for proposed lines through that part of the West.... Their support is what made the creation of Yellowstone National Park possible."[30] Condensing a large literature into one sentence, the historical record reveals how the Northern Pacific actively lobbied congressmen and funded studies to increase the chances that the bill to create the park would get passed. And all this before the company's line extended far enough west to make carrying passengers all the way to the park even possible: the park was established in 1872, but the Northern Pacific's line would not reach the area until 1883.

With the creation of the park came a debate over how it would be used. The *New York Times* suggested in early 1873 that "it is only necessary to render the Park easily accessible to make it the most popular summer resort in the country." But the author of that article was suspicious of such popularity, reminding readers of the negative effect that overcommodification wrought on earlier popular destinations, such as Niagara Falls. The article warned that that "hucksters" and "backmen" would overrun the area to sell their trinkets and must be banned to preserve the region's natural beauty. Allowing too much commercialization would be "fatal to the success of the Park as a place of resort for cultivated persons."[31]

Because of its near monopoly on access (one could reach the park by coach, but it was a long and arduous trek), the Northern Pacific would reap the economic rewards of delivering tourists to the park. This meant that *some* commercialization of the park was bound to arise. Old Faithful Inn, for example, would be built by the Yellowstone Park Association, itself tied to the Northern Pacific. But the Northern sought to ensure that the park's natural wonders would remain as unspoiled as it reasonably could. After all, the natural beauty and the region's outdoors activities were what the railroad's customers expected to find. Promotional material for the Northern Pacific focused on trout fishing as a key attraction besides the park's natural wonders. Protecting the park from overdevelopment made good business sense: if the allure of the region became spoiled, tourist interest would fade and so would revenues.

Yellowstone was not unique. Railroad companies also provided the impetus for the creation of the General Grant, Sequoia, and Yosemite national parks. And while the profit motive loomed large in driving their support for the preservation of large tracts of land, the railroads' backing was important to advance the nascent conservation and preservation movement. Even John Muir acknowledged at an 1892 Sierra Club meeting that "the soulless Southern Pacific R.R. Co., never counted on for anything good, helped nobly in pushing the bill for [Yosemite National Park] through Congress."[32]

What is important for my story is that by making access to the West and the newly created national parks possible, railroads fostered an appreciation among the public for nature and the activities that it offers. Advancing interest in tourism even helped fuel the formation of our national identity. Visiting the national parks—and this remains true even today—for many became something of a pilgrimage. Like religious shrines, places like Yellowstone or Yosemite provided "a physical location that embodies the values that orient the culture."[33] Visiting the national parks enhanced the public's growing belief in American uniqueness, both in its spirit and in its natural wonders.

As the national park system was being developed there arose conflicting forces concerning its use. Organizations like the Sierra Club pushed to preserve them in their purest form. The railroads that carried the tourists produced marketing strategies that were specific to the park they serviced. And because three separate departments (Agriculture, Interior, and War) of the federal government were assigned some degree of oversite for park use and management, no focused governmental agreement existed about what role the parks should play. "As various constituencies contended over the use and the meaning of the new parks," notes Marguerite S. Shaffer, "park advocates ranging from railroads interests to nature lovers began to seek some means of unifying park support and organizing park management."[34] Over the next couple of decades this disparity of interest converged with the creation of the National Park Service in 1916, a move that railroads endorsed.[35]

Part of this movement promoted the idea of making the parks open to all. Ken Burns observes in the preface to his 2009 book accompanying the documentary series *The National Parks: America's Best Idea* that "for the first time in human history, land—great sections of our natural landscape—was set aside, not for kings or noblemen or the very rich, but for everyone, for all time. . . . This idea, like our articulation of universal political freedom in the Declaration of Independence, should be so widely admired."[36] Add to this the fact that by the end of the century significant numbers of travelers—with increasing numbers coming from the middle class of America—were traversing the entire continent in a matter of days merely added to the perception of uniqueness.

Promotions to entice individuals for an adventure out West began in earnest with the opening of the transcontinental rail system west of the Missouri River. Even though in the early years the number of people visiting Yellowstone was limited—such travel was still relatively expensive and time consuming—demand for such experiences increased as the century wound to a close. Railroads sought to profit from carrying passengers to these destinations, but they also acted to protect these areas from overexploitation. With the seeds planted in the closing decades of the 1800s, the "See America First" movement would really take off with the widespread use of the automobile in the early decades of the twentieth century.

Railroads were preserving the landscape in other areas than just out west. Long before the national park idea took hold the land that would become Acadia National Park in Maine was owned by the Maine Central Railroad; that which became the Shenandoah National Park was popularized by several railroads; the Southern Railroad was taking visitors to the area that would become the Great Smoky Mountains National Park; and for almost fifty years prior to becoming a national park, a short-line railroad was shuttling passengers to and from the Hot Springs Reservation in Arkansas.[37] In each instance a railroad acted to preserve the landscape because the natural beauty is what drew passengers to take the train to these sites.

This is not to say that there was no environmental damage that occurred from laying tracks through forests and across prairies. My point is that railroads were much better custodians of the environment than most modern readers would imagine. "Simply, railroads brought leadership to the idea of preservation; commerce and the landscape could coexist," argues Alfred Runte.[38] And one must remember the context in which railroads operated. The public was much more willing to accept some environmental damage if the result was pushing the country farther along its path to supremacy, what most viewed as its manifest destiny. Railroads were a source of stewardship when it came to the environment. More to the point of this book, they also introduced an increasingly broad spectrum of the population to nature and outdoor pursuits.

A Focus on Fishing

I mentioned earlier that guidebooks were a popular form of advertising. Whether produced by a railroad or by another company, guidebooks became more than merely sources of mundane travel information. Once tracks reached Western sites, railroad guidebooks and pamphlets celebrated the transcontinental railroad as a manifestation of the opportunities of an "expanding republican empire."[39]

In addition to information about schedules and where to sleep and eat, many guidebooks provided the traveler with information on fishing and hunting resorts along a railroad's line. The title of one guidebook, *Guidebook to the Fishing and Hunting Resorts in the Vicinity of the Grand Trunk Railway of Canada*, makes it perfectly clear who the target audience was. This brochure even provided details regarding which different species of fish and game could be found where along the railroad's route. It also listed hotels and camps to stay in, guides available for hire, and what everything cost. Guidebooks such as this—and it is just one example of the many available—were published to attract not just the angler or hunter to ride on a particular railroad; they also succeeded in increasing interest in the outdoors by suggesting that entire family could enjoy the recreational activities, especially fishing, found at locations served by the railroad.

I mentioned that not all guidebooks were published by railroads. One example is *The Angler's Guide and Tourist's Gazetteer*, compiled by William C. Harris. Harris was the editor of the popular outdoors magazine the *American Angler* and an early and active advocate of developing a "fishing ethic" among the nation's growing number of anglers. This ethic basically amounted to the idea that an angler should keep only the fish one planned to consume. Today we refer to this as "catch-and-release" fishing.

Harris's guidebook included advertisements by railroad companies that proclaimed their ability to deliver the reader to best fishing locales. The 1886 edition of Harris's guidebook contained a number of such ads. One for the Maine Central Railroad declared it to be "the sportsman's line to the East." An ad for the Pennsylvania Railroad informed New York City residents they could enjoy "the best saltwater fishing on the New Jersey coast" simply by hopping aboard one of its trains. Because the railroad offered the fastest and comparatively least expensive (especially if one considered the value of time lost traveling) way to reach the "best" fishing locales, railroads increased the diversity of individuals wishing to fish waters celebrated by anglers from across the nation.[40]

Guidebooks and similarly enticing promotional materials were also produced by the railroads and made available through their ticket agents, usually at no expense. The Grand Rapids & Indiana Railroad suggested in one of its ads that appeared in all of the major sporting magazines that interested parties need only contact the general passenger agent in Grand Rapids, a Mr. A. B. Leet, who would gladly mail their "handsomely illustrated" 160-page guidebook to them. When the Northern Pacific's route was completed in September 1883, it wasted no time in promoting the wonders its passengers would find in the newly opened Northwest.[41] One of its promotional pamphlets was quite explicit about it: *6000 Miles Through Wonderland: Being a Description of the Marvelous Region Traversed by the Northern Pacific Railroad* was the title of one brochure. By the turn of the century this *Wonderland* pamphlet had become an annual publication with Yellowstone featured as the "Gem of Wonderland."[42]

The Chicago & North Western Railroad published one of the most inclusive railroad guidebooks. Its 140-page *North and West Illustrated: Facts for Tourist, Business & Pleasure Travel* covered everything a traveler could want. Its expansive subtitle says it all: *A Guide to the Lakes and Rivers, to the Plains and Mountains, to the Resorts of Birds, Game Animals, and Fishes; and Hints for the Commercial Traveler, the Theater Manager, the Land Hunter and the Emigrant.* All that for only twenty-five cents! Want to plan a fishing trip around a longer journey or a business trip? This guidebook made it easy, and many did.

Think of it: fishing for exotic wild trout in Colorado or the catching the exclusive American grayling found only in northern Michigan was made possible by taking the train. To say that hyperbole was the coin of the realm would be an understatement. In the brochure *Union Pacific Outings: Fishing in Colorado and Wyoming*, the text admonishes the desk-bound worker to leave the city and explore the wonders of a Western trout stream. Only there could one "enjoy the sky, the air, the mountains, the pines, the tackle, and the fish. It is the highest of the delicate art," as fly-fishing was often called. How to reach such a piscatorial paradise where "trout fishing is a symphony"?[43] Simple: buy a ticket on the Union Pacific.

All brochures were functional, proving time schedules, connecting trains, et cetera. Many were works of art. Even if one did not actually take that dream fishing trip, railroad brochures were often displayed like coffee table books: there to be thumbed through, shared with others. The Spokane & Inland Empire Railroad went so far as to include a sachet of fresh pine needles when it mailed one of its brochures, "to carry the balsa of the forests direct to the recipient." The Spokane & Inland did not scrimp on its brochures, either. The reason is because they recognized the marketing power of these publications. According to the company, "We often hear criticism about the expensive literature issued, but we insist that one folder that is artistic enough to be kept and shown or mailed to others covers more ground and makes a greater impression than twenty common, ordinary leaflets which are generally glanced over and thrown away. We have never yet seen one of our folders discarded on the floor of our cars or depots."[44]

I've covered only a small sample of the railroad guidebooks and brochures published in the latter part of the 1800s. I think you get the picture: railroads' promotional campaigns raised awareness among the public not only of what services they could offer, but also about the outdoors and the wonders that awaited the traveler. In all railroads made it easier than ever before for those interested to participate in this new recreation of sport fishing.

THE FISHING LINE

With railroad companies competing for the tourist-angler, their marketing campaigns used colorful and informative guidebooks to entice and inform potential customers. They also advertised their self-proclaimed invaluable service in national sporting magazines and newspapers. One example is the 10 August 1876 issue of *Forest and Stream*. (Indeed, looking through almost any issue from this and related outdoors publications from this time tells the same story.) It includes several railroad ads with similar self-serving claims. Calling itself "The Sportsmen's Route," a Chicago & Northwestern Railway ad tells the reader that fabulous fishing can be found at "a hundred points" along its tracks. Indeed, the Chicago & Northwestern claimed it was "unsurpassed by any" other railroad in the West in getting the interested angler to their desired destination. Another ad, this one for the Maine Central Railroad, states that it its tracks run through the "favorite haunts of the deer and trout." The Pennsylvania Railroad Company's ad averred that it serves all the "best localities for gunning and fishing" in Pennsylvania and New Jersey.

The Grand Rapids & Indiana Railroad ad that appears in this same issue promoted its ability to deliver you to Michigan trout streams, with the added inducement that it was their "aim to make sportsmen 'feel at home' on this route." As I have mentioned earlier, the company was so confident in this role that it nicknamed itself "the Fishing Line" in this and all of its promotional material. In what may be the most inspired marketing strategy to grab customers' attention was its use of a map showing the company's main lines and its principal connections across the eastern half of the United States. This was not unique: railroads often published such maps, either as stand-alone images or as covers to their pamphlets and brochures.[45] What makes the Grand Rapids & Indiana map stand out is its eye-catching image of an American grayling overlaying most of the map. It is so large that its tail begins in Nashville in the South, and its head points to where it can be found, in northern Michigan. The message is clear: anyone in St. Louis, Chicago, Detroit, Cincinnati, Philadelphia, or New York—any of the major cities with access the company's line—who wants to experience the thrill of catching this coveted fish need only to contact a Grand Rapids and Indiana ticket agent and book the trip.[46]

If the Grand Rapids & Indiana seemed to be the only line (it wasn't) to Michigan's famed trout waters, promotions by the Missouri Pacific Railway suggested that it in fact was the only line serving the fishing locales in Colorado (it wasn't). Calling itself the "Colorado Short Line," the company suggested that only it could deliver anglers and tourists to the best summer resorts and trout streams in the Rockies. Not only could railroads, unlike other available modes of transportation, deliver the intrepid outdoorsman and outdoorswoman to prized fishing locations around the country rapidly and safely, they would even transport your gear—fishing tackle, guns, and even dogs—at no additional charge.

Not interested in trout fishing? Other railroads were there to take you to wherever your fish of choice lived. In the December 1895–January 1896 issue of the *American Angler*, the Tropical Trunk Line informs readers that it can take you to "the finest tarpon fishing in the world." Another ad in that publication claims that anyone booking passage on the Mobile & Ohio Railroad will be transported to the finest fishing grounds in eastern Mississippi and southern Alabama. The ad copy cleverly asks, "Do you angle?" with the response "If so, angle with us." An ad for the Santa Fe Route simply declared that "the best hunting and fishing grounds of the United States" are accessible along its line.

Advertising campaigns like these extended well into the 1900s. I realize that this extends my self-imposed cutoff date a bit, but examples of railroad advertising that promoted place and fishing are worth the slight detour.[47] In a 1903 Rock Island System poster, a man is shown landing a large trout in a stream somewhere in Colorado, his happy family cheering him on from the bank. A 1904 ad for the Northwestern Pacific simply shows rising or jumping trout in a Yellowstone Park stream with the company's name and logo partially obscured by the leaping fish. Later, railroads adopted the minimalist visuals that became vogue during the 1920s and 1930s. One especially eye-catching poster used by the London & North Eastern Railway (LNER) in Britain is dominated by two trout swirling below a fly. Below the phrase "Try a Fly by the LNER," the company's contact information almost seems secondary to the visual. Such advertising promoted the glamour and genteel nature of fly-fishing. Of course, these characteristics were ones that the railroad also wanted you to associate with them. Many of the images also displayed women, often alone, enjoying their time in the water participating in this "delicate art." By showing women fishing, railroads helped to broaden the appeal of fishing and thus influenced the broadening of the sport fishing movement. Of course, they also were meant to encourage women to ride the train.[48]

These promotional materials—magazine ads, maps, and posters—exploited the railroad industry's absolute advantage in travel. Providing low-cost fares with practical travel times, and by using imagery that promoted fishing, railroads fueled the public's interest in travel and in sport fishing.

REPURPOSING OLD LINES

Many railroads realized that capturing a larger share of the growing sport fishing and tourist market took more than just advertising. To do this some railroads redirected resources into the expanding tourist trade. Consider the Jackson, Lansing & Saginaw Railroad in Michigan. With track running up through the middle of Michigan's Lower Peninsula, the company was built originally to carry logs from the state's northern pine forests to sawmills to the south. There the logs were transformed into lumber and sent south to meet the burgeoning construction demands in cities like Chicago and Detroit. As the forests were cleared, there were fewer logs to ship. As revenues from this line of business began to decline, the Jackson sought alternative sources of revenue. One of these was the nascent tourist industry. To supplement its revenues, the railroad "began advertising hunting and fishing trips to the Grayling area, luring increasing numbers of sportsmen to the game and fish available in the newly opened region."[49] By promoting fishing in the northern areas of Michigan's Lower Peninsula, the Jackson railroad heightened the popularity of fishing the area's increasingly famous rivers, such as the AuSable River. As the railroads improved access to these once inaccessible areas, the emerging hospitality industry of lodging, food service, and fishing guides for visiting anglers offered much needed income to those who had relied on the forest industry for their livelihoods. And, too, it raised awareness of and participation in sport fishing.

Other Michigan railroads capitalized on the growing popularity of tourists and anglers looking to sample this increasingly popular "destination" of northern Michigan. The Detroit, Lansing & Northern Railroad, which also started by transporting logs out of the northern forests, switched to serving the tourist trade as the logging industry in Michigan faded. Its ads encouraged men to "take your wife and some fishing tackle and go north."[50] In 1883 the guidebook *Detroit and the Pleasure Resorts of Northern Michigan* offered to take the "pleasure seeker" to "Attractions Unsurpassed in this Country." To the angler it promised access to the "Jordan and Boardman Rivers, Famous for Trout Fishing," and to the "Manistee River, the Celebrated Grayling Stream."[51] With such fishing meccas only a short train ride away, many from Detroit, Chicago and other nearby metropolitan areas tried their hand at sport fishing. And, with easy connections, anglers across the eastern half of the country began frequenting these increasingly popular streams.

Not all railroads set their sights on luring anglers from a national market. The Salem & Little Rock railroad in Missouri offers one such example.[52] Completed in 1873, it hauled iron ore from mines near the town of Salem, Missouri, to St. Louis, about 120 miles to the northeast. As sport fishing became increasingly popular, the "Salem Line," as it was known, began offering passenger service to those wishing to fish the spring-fed Current River, which *Forest and Stream* magazine recognized as "one of the best fishing streams" in the country.[53]

Most of the Salem Line's passengers came from St. Louis, Missouri, to the east, or Springfield, Missouri, to the west. In both instances passengers would first take the Frisco railroad to Cuba, Missouri, and then transfer to the Salem Line, which took them south to Salem. From there wagon service was available to reach the Current River. For many it was an easy day trip, thus reducing the up-front costs of an outing on the river. With much improved accessibility, the Current River became one of the most fished streams in the state. Even though the Current was known for its variety of fish, especially small-mouth bass, beginning in the early 1890s it was often stocked with trout by federal and state fish commissions. To ensure interest in fishing the Current, there is evidence that the Salem Line also stocked the river with trout to guarantee the availability of fish for its passengers to catch.[54]

THE FISHING TRAIN

Some railroads offered "fishing trains"—or "excursion trains," as they were sometimes called—to meet the growing interest in sport fishing. For a special low fare, these specialized trains enabled (mostly) urban anglers to escape the city and visit nearby streams for a day of fishing. Fishing trains departed from cities like Denver, Butte, and Great Falls, dropping off anglers and their gear at stops along famous trout streams, such as the Big Hole and Beaverhead Rivers. After a day on the water, these anglers, along with their catch, would wait at a nearby rail stop and flag down the train heading back to town.[55] The Maine Central Railroad offered a similar arrangement. An angler could hop aboard one of its fishing trains at any one of many stops and, upon notifying the conductor, disembark at a chosen fishing spot. Arrangements to pick up the anglers thus distributed along the tracks on the return trip were made with the conductor. These are only two of many examples of how railroads expanded the opportunities for anglers who wished to enjoy a relatively inexpensive day's worth of fishing.

Some railroad companies promoted fishing (and ridership) by conducting special promotions specifically for anglers. The *Transactions of the American Fisheries Society* reports that the Denver & Rio Grande Railroad offered a twenty-dollar gold piece to any *fly fisherman* (note the exclusivity!) who landed a rainbow trout that weighed ten pounds or more—verified of course—from any river along their line. It is reported that the railroad paid up about twice a year, and that the lunkers usually came from Colorado's Gunnison River.[56] Promotional programs such as these (fishing trains and prize gimmicks) were all done to build ridership, but they also induced more people to get out and go fishing.

THE SPECIALIZED CAR

Beginning in the 1880s, some railroad companies recognized the trending interest in all things outdoors and began renting private train cars specifically fitted up for hunting and fishing parties. The special cars of the Northern Pacific, for example, came complete with a cook and porter. If you were planning an extended hunting or fishing trip, the specialized cars could be parked on any Northern Pacific's right-of-way for a month at a time. The Pullman Company also offered two special cars that parties of hunters or fishermen could rent. The *Davy Crockett* and the *Izaak Walton* were rolling hunting lodges, with amenities similar to that of the Northern Pacific's cars. The rental fee was thirty-five dollars per day, a little over one thousand dollars in modern terms. Once you realize that this was a flat fee, and the capacity was ten passengers, the daily per passenger cost is quite reasonable for what you got.[57] The Pullman's rental fee, like that for one of the Northern Pacific cars, not only included a cook and waiter but, for hunters, included kennels for your dogs.

STOCKING STREAMS

The historical record reviewed in chapter 4 indicates that hundreds of railroads actively assisted the federal and state fish commissions in their attempts to manage the quantity and diversity of the nation's fishery. It turns out that many railroads also enthusiastically stocked the rivers and streams along their tracks.

The Frisco's stocking of streams and lakes near its Lodge in Missouri is an example where a direct benefit accrued both to the railroad and to the public. Fish have a tendency to move around, so those stocked in the streams on the Frisco's property migrated to publicly accessible parts of the stream, and into the Gasconade River. More often than not, however, railroads simply stocked rivers and streams along which their track ran. I've already noted that the Salem Line in Missouri stocked the Current River with trout to attract passenger-anglers. And I've noted before that stopping a train so fish messengers could stock a nearby stream was a common occurrence.

Out west the Northern Pacific Railroad worked to increase the diversity of the fish population in Lake Pend Oreille to attract more passengers. Their argument was that if more species of fish were available it would increase the lake's appeal to tourists and anglers and thus boost ridership. The record also shows that representatives for the Denver & Rio Grande Railway applied for and received annual allotments of cutthroat and brook trout from the Commission that were then deposited by the railroad into the Provo River in Utah, which flowed near its tracks.[58] Appeals by railroads to the Commission, and undoubtedly to state commissions as well, strengthens the argument that railroads often acted to maintain or increase the population of trout and other species in waters near their tracks. Yes, such intervention was driven by a profit motive: more fish (hopefully) equals more passenger-anglers. But these actions also helped increase the public's interest and participation in fishing.

While some railroads stocked fish to attract anglers to ride their trains, others went as far as to repurpose their rolling stock to support such endeavors. The best-known example of this is the Flint & Pere Marquette Railway (later the Pere Marquette Railroad) in Michigan. In 1873 the company built an inspection locomotive, a train car that looks like a small locomotive with a small passenger car appended to it. These specialized cars were used to travel along the tracks checking for problems, such as loose rails or washouts. The story goes that the superintendent of the railroad, Sanford Keeler, used this inspection car, which he nicknamed "the Peggy," for personal fishing trips in Michigan and to carry cans of trout fry that were stocked in streams along the railroad's right-of-ways.[59]

My focus has been on trout, but there is evidence that railroads also took an active role in transplanting other species of fish on their own. We know, for example, that smallmouth bass were first introduced in Maryland in 1854, when a B&O employee transported fish from the Wheeling River in West Virginia and deposited them into the C&O Canal Basin in Cumberland, Maryland. During the carp craze a number of railroads in Texas (and to be sure other states as well) took delivery of carp from the Commission to stock their own lakes and ponds.

Let me clarify the importance of the foregoing examples. Yes, the railroads engaged in these activities in order to increase ridership. Trout or whatever fish was planted in streams and lakes near the railroad's tracks attracted anglers to ride their train. But the railroads could not preclude anyone from trying their hand at catching the fish if they did not take the train. Stated differently, the railroads created a public good: in most instances they could not exclude nonriders from catching the fish they had planted. Because railroads generally could not exclude anyone from fishing in the rivers or lakes, railroads thus created a positive externality for anglers and the public at large. Why the general public? Remember, at the heart of the restocking movement was to increase the stock of food fish in the country. Through these acts of profit maximization the railroads enhanced the public's chance to catch fish as a food source and broadened their interest in fishing as a recreational activity. Railroads thus were instrumental in advancing sport fishing in America.

A Final Note

Railroads democratized sport fishing in America. By the late 1800s, the angler—man *and* woman—from nearly anywhere in the country could travel by rail for only a few days to experience the thrill of fishing in streams that had once been the exclusive haunts of the wealthy. Urban residents across income levels now found it easier to take up fishing, whether it was to upstate New York or a vacation to Colorado or Michigan. More could enjoy a day trip to a nearby stream. Fishing stories in popular magazines regaled readers with stories of such trips, often spending as much space writing about the train ride as the actual fishing. Railroads were instrumental in meeting, and I would argue causing, the increased demand for travel in general and a new, emerging idea of tourism.

Through a variety of means, railroads encouraged the public's growing attraction to sport fishing as a recreational pastime. I think it is reasonable to argue that the development of sport fishing in the United States would not have occurred at the pace it did without the railroads' involvement. To put a finer edge on that idea, the next chapter will show how railroads helped participation in sport of fishing grow by making it more inclusive.

COMPARATIVE TIME-TABLE,
SHOWING THE TIME AT THE PRINCIPAL CITIES OF THE UNITED STATES, COMPARED WITH NOON AT WASHINGTON, D. C.

There is no "Standard Railroad Time" in the United States or Canada; but each railroad company adopts independently the time of its own locality, or of that place at which its principal office is situated. The inconvenience of such a system, if system it can be called, must be apparent to all, but is most annoying to persons strangers to the fact. From this cause many miscalculations and misconnections have arisen, which not unfrequently have been of serious consequence to individuals, and have, as a matter of course, brought into disrepute all Railroad-Guides, which of necessity give the local times. In order to relieve, in some degree, this anomaly in American railroading, we present the following table of local time, compared with that of Washington, D. C.

NOON AT WASHINGTON, D. C.	NOON AT WASHINGTON, D. C.	NOON AT WASHINGTON, D. C.
Albany, N. Y......12 14 P.M.	Indianapolis, Ind...11 26 A.M.	Philadelphia, Pa....12 08 P.M.
Augusta Ga.......11 41 A.M.	Jackson, Miss......11 08 "	Pittsburg, Pa......11 48 A.M.
Augusta, Me. .....11 31 "	Jefferson, Mo......11 00 "	Plattsburg, N. Y..12 15 P.M.
Baltimore, Md.....12 02 P.M.	Kingston, Can.....12 02 P.M.	Portland, Me......12 28 "
Beaufort, S. C.....11 47 A.M.	Knoxville, Tenn....11 33 A.M.	Portsmouth, N. H.12 25 "
Boston, Mass......12 24 P.M.	Lancaster, Pa.....12 03 P.M.	Pra. du Chien, Wis.11 04 A.M.
Bridgeport, Ct.....12 16 "	Lexington, Ky.....11 31 A.M.	Providence, R. I...12 23 P.M.
Buffalo, N. Y......11 53 A.M.	Little Rock, Ark...11 00 "	Quebec, Can......12 23 "
Burlington, N. J...12 09 P.M.	Louisville, Ky.....11 26 "	Racine, Wis.......11 18 A.M.
Burlington, Vt....12 16 "	Lowell, Mass......12 23 P.M.	Raleigh, N. C.....11 53 "
Canandaigua, N.Y.11 59 A.M.	Lynchburg, Va.....11 51 A.M.	Richmond, Va.....11 58 "
Charleston, S. C...11 49 "	Middletown, Ct....12 18 P.M.	Rochester, N. Y...11 57 "
Chicago, Ill......11 18 "	Milledgeville, Ga...11 35 A.M.	Sacketts H'bor, NY.12 05 P.M.
Cincinnati, O......11 31 "	Milwaukee, Wis....11 17 A.M.	St. Anthony Falls..10 56 A.M.
Columbia, S. C....11 44 "	Mobile, Ala.......11 16 "	St. Augustine, Fla.11 42 "
Columbus, O......11 36 "	Montpelier, Vt....12 18 P.M.	St. Louis, Mo.....11 07 "
Concord, N. H.....12 23 P.M.	Montreal, Can.....12 14 "	St. Paul, Min......10 56 "
Dayton, O........11 32 A.M.	Nashville, Tenn....11 21 A.M.	Sacramento, Cal...9 02 "
Detroit, Mich.....11 36 "	Natchez, Miss.....11 03 "	Salem, Mass......12 26 P.M.
Dover, Del.......12 06 P.M.	Newark, N. J......12 11 P.M.	Savannah, Ga.....11 44 A.M.
Dover, N. H......12 37 "	New Bedford, Mass.12 25 "	Springfield, Mass...12 18 P.M.
Eastport, Me......12 41 "	Newburg, N. Y....12 12 "	Tallahassee, Fla...11 30 A.M.
Frankfort, Ky.....11 30 A.M.	Newburyport, Ms..12 25 "	Toronto, Can......11 51 "
Frederick, Md.....11 59 "	Newcastle, Del....12 06 "	Trenton, N. J....,12 10 P.M.
Fredericksburg, Va.11 58 "	New Haven, Conn..12 17 "	Troy, N. Y.......12 14 "
Frederickton, N. Y.12 42 P.M.	New London, " ..12 20 "	Tuscaloosa, Ala....11 18 A.M.
Galveston, Texas ..10 49 A.M.	New Orleans, La...11 08 A.M.	Utica, N. Y......12 08 P.M.
Gloucester, Mass..12 26 P.M.	Newport, R. I.....12 23 P.M.	Vandalia, Ill......11 18 A.M.
Greenfield, " ..12 18 "	New York, N. Y...12 12 "	Vincennes, Ind....11 19 "
Hagerstown, Md...11 58 A.M.	Norfolk, Va.......12 03 "	Wheeling, Va.....11 45 "
Halifax, N. S......12 54 P.M.	Northampton, Ms..12 18 "	Wilmington, Del...12 06 P.M.
Harrisburg, Pa....12 01 "	Norwich, Ct......12 20 "	Wilmington, N. C..11 56 A.M.
Hartford, Ct.....12 18 "	Pensacola, Fla....11 20 A.M.	Worcester, Mass...12 21 P.M.
Huntsville, Ala....11 21 A.M.	Petersburg, Va....11 59 "	York, Pa.........12 02 "

By an easy calculation, the difference in time between the several places above named may be ascertained. Thus, for instance, the difference of time between New York and Cincinnati may be ascertained by simple comparison, that of the first having the Washington noon at 12 12 P. M., and of the latter at 11 31 A. M.; and hence the difference is 43 minutes, or, in other words, the noon at New York will be 11.17 A. M. at Cincinnati, and the noon at Cincinnati will be 12 43 P. M. at New York. Remember that places *West* are "slower" in time than those *East.* and *vice versa*.

FIG. 1. Until the 1870s railroads used different time zones. Shown above is the time table in *Travelers Official Rail Way Guide* for June 1868. Passengers used these time tables to match schedules from different railroads using noon in Washington DC as the basis of comparions. Author's collection.

FIG. 2. The completion of the first transcontinental railroad was marked with the driving home of the golden spike at Promontory Summit, Utah Territory, 10 May 1869. The Central Pacific Railroad, on the left, met with the Union Pacific Railroad, on the right. The transcontinental railroad linked Sacramento in the West to Omaha in the East and made rail travel across the country possible. Photo by Andrew J. Russell. Adam Cuerden, Yale University Libraries.

FIG. 3. Seth Green (1817–88) was a pioneer in American fish culture. He is credited with numerous innovations that led to marked increases in fish production, and to his numerous experiments with fish hybridization. His impact was so notable that he is often referred to as the "father of U.S. Fish culture." Seth Green, *Trout Culture* (1870).

FIG. 4. Spencer Fullerton Baird (1823–87) was a noted naturalist with the Smithsonian Institute and the first U.S. fish commissioner. He helped establish the U.S. Fish Commission as a major contributor to the scientific study of aquatic animals and plant life. Under his direction the Commission undertook many significant experiments in using fish culture to protect the nation's fishery. Photo by William Bell. Smithsonian Institution Archives, Record Unit 95, Box 2, Folder 6.

FIG. 5. Robert Barnwell Roosevelt (1829–1906) was a key activist in the early use of fish culture to combat the over-harvest of the nation's fish stock. As a U.S. congressman he is credited with promoting the early activities of the U.S. Fish Commission. He advocated the use of conservationist measures to combat overfishing and the decimation of the nation's fishery. Library of Congress, Brady-Handy Collection, LC-DIG-cwpbh-00644.

FIG. 6. Livingston Stone (1836–1913) was an early American fish culturist. He is best known for his early work with Atlantic salmon and as the manager of the Fish Commission's salmon and rainbow trout hatcheries in Northern California during the 1870s. Anders Halverson (https://andershalverson.com/u-s-fish-commission/).

FIG. 7. This image shows the salmon station on the McCloud River in Northern California. The hatching house is shown in the center of the picture, flanked by Mount Persephone. The structure in the foreground is the current wheel, used to divert water for use in the hatchery. At the time this picture was taken it was known as the Baird Station. *Bulletin of the United States Fish Commission*, 1896.

FIG. 8. Fish fry were transported in standard milk cans. The cans were sent from the hatchery by train and upon reaching their destination the fry were deposited in streams and lakes. Lovells Township Historical Museum.

FIG. 9. Beginning in 1881 the U.S. Fish Commission built a fleet of railroad cars dedicated to transporting fish. The Commission used these "fish cars" to freely deliver fish to state commissions and individuals across the country. Shown here is a fish car parked outside the U.S. Fish Commission building with workers loading milk cans that will be transported by the fish car. Smithsonian Institute Archives. Image 95–1179.

FIG. 10. The interior of a fish car. This illustration shows the fold-down bunks that Commission employees, known as "fish messengers," used for sleeping. On each side of the aisle are compartments that held cans full of fish. When covered, chairs were placed on them. Most fish cars were fitted with kitchens and other amenities that made long-distance travel possible. Freshwater and Marine Image Bank, University of Washington. https://digital-collections.lib.washinton.edu/digital/collection/fishimages/id/35220/rec/208.

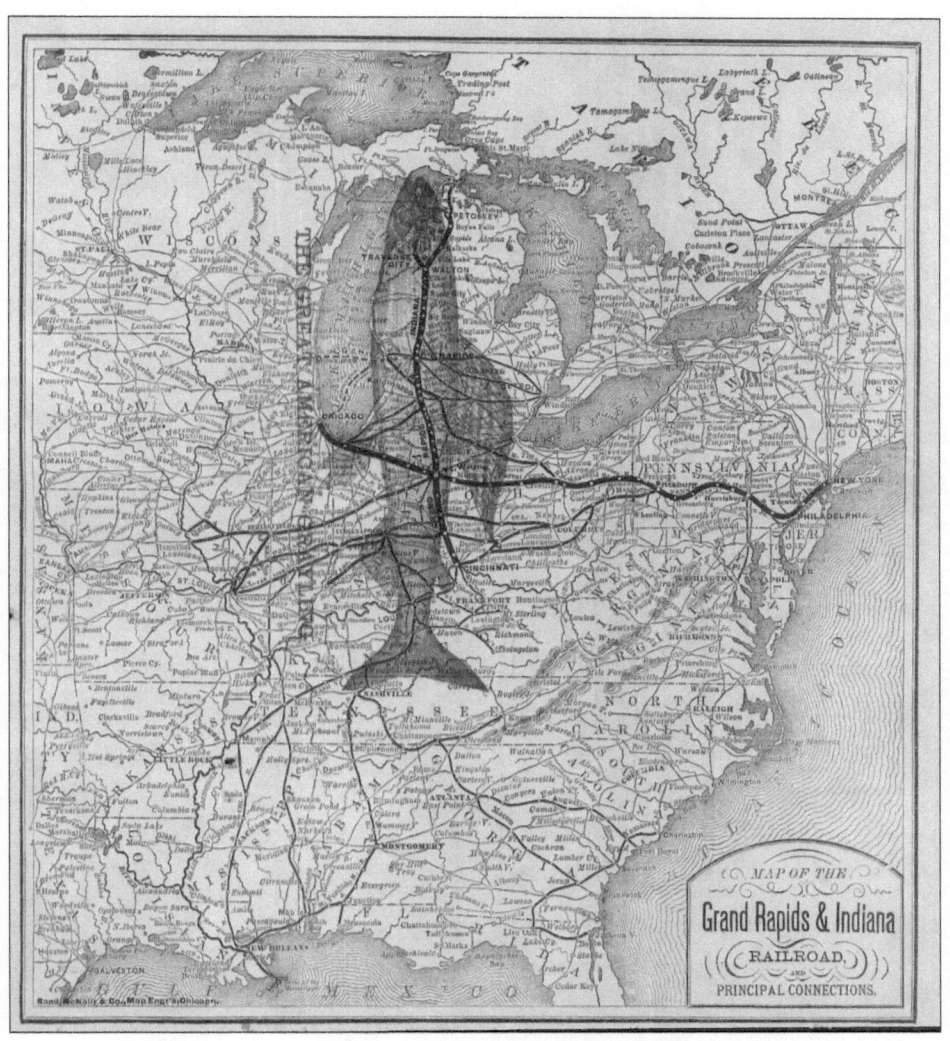

FIG. 11. Calling itself "the Fishing Line," the Grand Rapids & Indiana Railroad used images like this to attract customers. The map shows the railroads that connected to its main line running north–south through Michigan, and what an angler could find by taking this line: the Great American Grayling, found only in the northern streams of Michigan. Archives of Michigan.

FIG. 12. Railroads often touted their ability to transport individuals to the "best" vacation destination. This Chicago, Milwaukee & St. Paul Railway poster illustrates such a marketing strategy. By taking this railroad, the "sportsman's favorite," the avid sportsman or the neophyte will be delivered to the best hunting and fishing destinations. In addition, along the journey passengers will experience the "finest" scenery. Author's collection.

FIG. 13. The Lackawanna Railroad featured the fictional Phoebe Snow in its advertisements. Always clad in a pure white dress, the young Miss Snow is often shown riding the train alone. In this ad she not only rides in comfort and safety, but also enjoys a beautiful countryside, complete with a flowing stream. Railroad Museum of Pennsylvania.

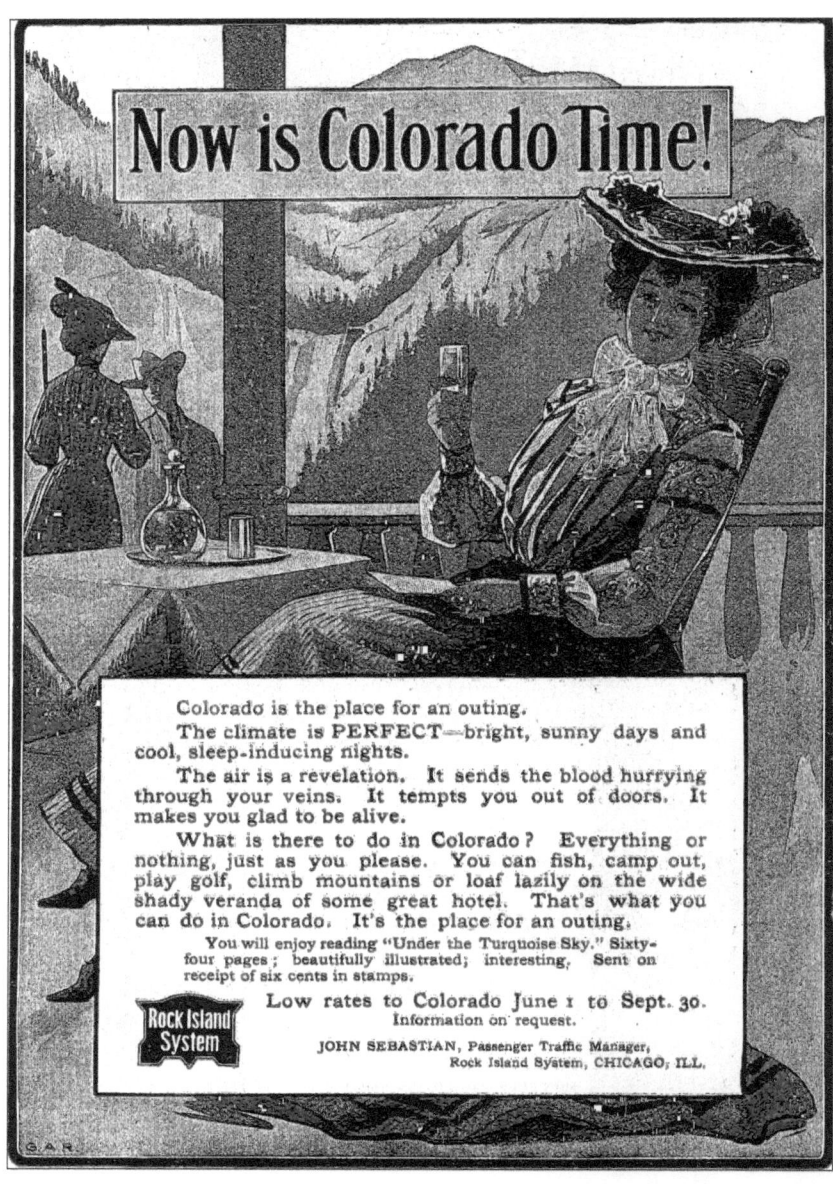

FIG. 14. Railroads often featured women in their ad campaigns. This *Cosmopolitan* magazine ad for the Rock Island Railroad conveys the idea that Colorado is a destination vacation where women can enjoy a variety of experiences, including fishing or camping. To reach this wonderland, take the Rock Island Railroad. Author's collection.

FIG. 15. William Welsh produced a number of posters for the Pullman Company. This poster is one of several featuring stylish young women in a variety of outdoor settings. The Pullman posters encouraged travel and outdoor activities with lines like "Vacation lands are calling." It also reminded passengers that if they traveled on a Pullman car, they would be assured of a safe and comfortable journey. Photo by Fred Ash. Author's collection.

RAILROADS, WOMEN, AND FISHING

In the previous chapter, I explored the various ways in which railroads acted, both directly and indirectly, to advance the public's awareness of the outdoors and its recreational activities. Among these activities was the increasingly popular pastime of fishing as a sport rather than a means to put food on the table. I mentioned in that chapter that railroads actively sought to increase female ridership through a number of targeted marketing campaigns, many of which linked ridership to the outdoors.

This chapter takes a deeper dive into railroads' influence on sport fishing. My focus here is specifically on their attempt to attract female passengers and show how this in turn helped expand the number of those interested in sport fishing. The message of their marketing campaigns was clear: women could travel by train, even long distances, in comfort and, perhaps more importantly, in safety. Lowering this barrier to travel liberated women from Victorian norms. Railroads pushed the idea that a woman could experience the world, including travel to outdoor locales, without a man by her side. As the century drew to a close, travel and tourism by women became an increasingly popular pastime. And, by generating interest among women in travel, railroads simultaneously promoted the idea of the woman as angler, the unintended consequence of which was the growing participation of women in the environmental and conservation movement.[1]

Improving Service

I noted in chapter 1 that there were several important technical innovations introduced by railroads during the latter part of the century—for example, Westinghouse brake systems, better couplers, and a standardized rail gauge, among others. Other improvements also were made that stemmed from competitive forces at work. These changes had more to do with improving service. Together the railroads strived to make train travel safer and more pleasurable.

THE SLEEPER CAR

Most passengers in the early years traveled in day coaches: train cars with rows of benches with some overhead storage. Longer trips required overnight stays, which required passengers to seek lodging in hotels and inns at stops along the route. Segmenting a trip with such stops translated into traveling for more time than some wished, and having to stay in sometimes dodgy lodgings was not the most effective inducement to take the train, especially for women and children. The sleeper car is one example of an innovation that improved the service offered to passengers.

The original sleeper car was one in which daytime seating converted into a bed for the night: a fold-down sleeper berth was eminently better than sleeping in one's seat. The sleeper car meant that riders had more privacy than those in the day coaches and offered a better travel experience, at least for those who were willing to pay for it. Theodore T. Woodruff is credited with developing the first sleeper car in 1857, but others quickly followed. Within a year the New York Sleeper Car Company and the Case Sleeping Car Company produced sleeper cars for different railroad companies. Among the many sleeper car manufacturers a Chicago company would rise above all.

George Pullman built its first sleeper car in 1859 at the Chicago & Alton Railroad shop in Bloomington, Illinois.[2] Pullman's sleeper cars were a marked improvement, offering more privacy, amenities, and better sleeping conditions. Pullman set up his own company in 1862 to fill the demand from some of the traveling public for high-end sleeper cars. The Pullman sleeper car quickly became the gold standard of the industry. They offered luxury accommodations—carpeting, tables, upholstered chairs, and, of course, sleeping berths—and heretofore unparalleled customer service. Within a short time, the Pullman Palace Car Company achieved such a reputation for its product that it received financial backing from notables such as steel magnate Andrew Carnegie and Chicago retailer Marshall Field. By the end of the century, Pullman built a near monopoly on the sleeper car business.

The Pullman palace car will play an important role in marketing train travel to women. As I will discuss, the Pullman Company's ads and service to back them up helped ease any qualms a woman might have about traveling by train.

Because the Pullman car offered a comfortable and safe way to travel, advertising suggested that a woman and her family traveling in a Pullman car would think they were sitting at home in their parlor instead of riding on a train.

By the 1880s Pullman cars and the railroads who moved them recognized that the tourist market was much broader than only the higher-income traveler. To capture a broader market share, Pullman introduced its tourist car. Even though it retained much of the allure that Pullman had established for its palace car, the tourist-grade cars were neither luxurious nor exclusive. But because they offered the customer something better than the standard coach "the tourist car made Western pleasure-travel possible and reasonably comfortable for many who could not stand the bare wooden benches of the earlier second-class sleepers. . . . Now a porter made the berths." This innovation expanded the market for tourists by making long trips more appealing. Service was paramount, even to those not ensconced in the luxury of the first-class cars. Indeed, "on some trains a special tour conductor rode the cars from coast to coast, giving a cut-rate approximation of what the first-class excursionists enjoyed, and at no extra cost."[3] The introduction of the new Pullman tourist car reduced the discomfort of long distance travel for all, but especially for women: now, instead of sleeping on the bench seat in the coach car, more and more women and their families could enjoy the comfort of a Pullman.

THE DINING CAR

The Chicago & Alton Railroad offered the first dining car, naming it *Delmonico* after the well-known New York restaurant. The idea behind the dining car is simple: passengers could now eat and drink as the train whizzed along, avoiding the mad rush to pick up something at brief stops at stations along the route where the quality of the food was sometimes questionable, and the prices were often high. "The demand for rapid traveling," noted *Harper's Weekly* in August 1868, "has led to the abandonment of many stops at lunchrooms and stations. . . . Instead [with dining cars] the passenger enjoys his soup, his fish, his joint, Clicquot [champagne] and coffee while the train runs fifty miles an hour."[4]

To meet this competition, Pullman went one step further and brought out its hotel car. It offered the passenger two services in one car: dining and overnight sleeping facilities. For those who could afford it, the hotel car offered the most accommodating and comfortable experience that existed in long-distance travel by land. As you will see, both the dining car and the hotel car were important parts in the railroads' promotion of long-distance travel to women.

The public loved the dining car. Even so, many railroads refused to include them on their trains. One explanation is that the cost of operating a dining car was not trivial. In addition, most passengers were traveling in a class that did not offer access to the dining car. The fact that providing a dining car could be detrimental to the bottom line explains why in 1881 three long-distance carriers—the Chicago, Burlington & Quincy Railroad Company; the Atchison, Topeka & Santa Fe Railroad Company; and the Union Pacific Railroad Company agreed to not carry dining cars on their trains without first notifying the other signatories to this noncompetitive agreement. The pact was broken, however, when the Northern Pacific Railroad Company (not part of the alliance) made dining cars part of its regular service on its newly completed route between Duluth, Minnesota, and Portland, Oregon, in 1883. Given the intense competition for passengers that existed between rival railroads, dining cars soon became common amenities on all of the major carriers.

HARVEY GIRLS

Though not an innovation in rail travel in a physical sense, the so-called Harvey Girls played a significant role in popularizing travel to the West. With the Northern Pacific promoting its ability to deliver you to Yellowstone Park, the Atchison, Topeka & Santa Fe Railroad marketed the Southwest as an equally fascinating destination. At the time a significant number of railroad passengers still carried on board their own lunches or other forms of sustenance. If one did not carry a meal and did not have access to a dining car, it typically was a mad dash into the depot, if there was one, or into town to find food. There was no time to be picky, since stops often lasted twenty minutes or less. Fred Harvey saw a way to solve this problem.

Harvey opened two eating establishments on the Kansas Pacific Railroad line, one in Wallace, Kansas, and the other in Hugo, Colorado. Harvey showed with his two cafés that he could provide the traveler with decent food at reasonable prices. Equally important, he could do so quickly. The management of the Atchison, Topeka & Santa Fe Railroad realized that one way to attract passengers, other than seeing the sights of the Southwest instead of the Northwest, was to partner with Harvey and expand his network of depot cafés. So, in 1889, the Santa Fe and Harvey agreed to an exclusive contract to build and manage a series of his cafés along their routes west of the Missouri River. The popularity of these establishments and the fast-spreading reputation for proficient and friendly waitresses—the Harvey Girls—was a boon for the railroad. By the early 1900s, the simple diner concept had grown into a series of hotels and eating establishments, known as "Harvey Houses," that created the kind of tourism infrastructure that would help establish the Southwest as a destination, and the Atchison, Topeka & Santa Fe Railroad as the railroad of choice.[5]

So how is all of this related to encouraging women to travel by train? The comment that many brought box lunches translates into women being given that chore when traveling with husbands and families. The rough and tumble to get off the train and back in order to get something to eat was not appealing to most women (nor to many men), especially if they were traveling with children. The slogan "Meals by Harvey" meant that families could enjoy a delicious, reasonably priced meal and be cared for in style along their journey. The reputation of the eating establishments was such that already by 1890 the Atchison, Topeka & Santa Fe Railroad ran an ad in the *Topeka Daily Capital* stating that "everyone knows what kind of meal Fred Harvey Gets up," a drawing card for riding on their railroad over the competition.[6]

These changes were important innovations that improved all traveler's experience, but especially women. I've mentioned that train travel, especially long-distance, was still relatively new. Articles found in magazines about a fishing trip to Michigan or Colorado often had as much to say about the train ride itself as it did the fishing. Anything that improved the experience of traveling induced more people to travel by train. In addition to improving the travel experience for all passengers, railroads also undertook other changes that were meant to encourage a still largely untapped segment of the market—women—to travel by train.

Train Travel for Women

Men would always account for most of the passenger traffic. The quote from *Harper's Weekly* attests to who was considered the representative passenger: *he* enjoys *his* soup, *his* fish, *his* champagne, and *his* smoke. But Victorian society was undergoing a significant transformation in the latter part of the nineteenth century, and part of that transformation was rethinking what constituted socially acceptable behavior for women. As more and more women were throwing off the constraints of Victorian society, they began to explore the world around them. Railroad travel offered an avenue to do this.

Railroads were a catalyst in this societal transformation. Whether it was done deliberately or not I cannot say. Regardless of their motives, other than to increase revenues, railroad advertising aimed to convince women—and, arguably, the concern of men for their wives, daughters, or sisters—that traveling by train was comfortable and, more important, safe. The evidence for such a claim is the fact that after the Civil War railroad advertisements often featured a well-appointed young woman riding in a train car happily enjoying the always picturesque landscape as it passes by. The images exude a sense of security: there is no apparent anxiety in the female passenger. Such advertising implanted in the observer the idea that women could (should!) expect a setting of "public domesticity" when traveling by train. Railroads "attempted to bring the cultural associations and behaviors of home life to bear upon social interactions among strangers, to regulate public interactions and delineate the boundary of Victorian respectability."[7] The idea was to convince women (and the men who "allowed" them to travel) that the surroundings of a train car, especially a Pullman car, was akin to that of entertaining in their own homes.[8]

This push to convince women of the safety and ease of train travel is evidenced in many ways. A travel brochure produced by the Chicago & North Western Railroad Company left nothing to doubt: "Between Chicago and several terminal stations, it [the Chicago & North Western Railroad Company] is the only road [on which] *which women and children can have a room to themselves, and be as isolated as in their homes*" (emphasis added).[9] Pullman ads in the latter decades of the 1800s similarly evoked such safety. Even though the majority of passengers were men on business trips, more often than not a Pullman advertising poster showed a single woman enjoying the comfort and safety afforded by riding in one of their cars. By convincing women that train travel was as safe and comfortable as their own homes, more women were encouraged to take to the rails. Their adventures increasingly included outdoor adventures, like fishing. Fishing was not only considered an acceptable sport for women, but with increased rail travel it would also become an increasingly popular pastime.

THE LADIES CAR

For the woman traveler who could not afford the luxury of a Pullman palace car, many railroads offered a "ladies car." The concept of a ladies car had been around for a number of years. The Cumberland Valley Railroad Company provided a separate compartment for women in some of its cars as early as the mid-1830s. The Philadelphia, Wilmington & Baltimore Railroad offered separate cars for women and children. As train travel increased in popularity among women, the availability of the ladies car became more typical.

Providing a ladies car was not without controversy, however. In 1874 a male passenger tried to ride in one of the Chicago & Northwestern's ladies car, and, when he refused to move, he was put off the train at a station near the town of Oshkosh, Wisconsin. He sued the railroad for this abuse. The case made it all the way to the Supreme Court of Wisconsin, which ruled that the railroad had the right to create a separate space for women passengers. Because the railroad advertised that it would provide a safe environment for women traveling on their line, the court argued that the company had the moral obligation to fulfill its promise and protect them.

A similar event occurred in 1877, when a male passenger tried to board a New York Central & Hudson River Railroad ladies car at one of its stations. After a minor ruckus, he too was forcibly removed by the brakeman. He also sued the railroad. And once again the New York state courts ruled that the railroad company had the legal right to reserve a car specifically for women traveling alone or with children. This case also gained notoriety, enough that the court's decision was covered by the *New York Times* in 1 February 1879 article titled "The Ladies' Car." While all of this may seem sexist through today's lens, it indicates just how serious the railroads were when they advertised that women could expect to ride in safety. It helped business, and it helped women shed the shackles of Victorian constraints.

In addition to having separate cars for women, some railroads instituted special instructions for the treatment of women passengers by company employees. The personnel guide circulated by the Philadelphia & Reading Railway Company is illustrative of the culture that railroads looked to maintain on its trains and in its stations. The guide states that the railroad's employees (mostly men) "must be respectful and courteous to all passengers, *especially women traveling alone*" (emphasis added). An important part of their job was to provide "polite attention to their [the women's] requests and all desired information as to routes, baggage or connections." Lest someone get the wrong idea, the employees were warned to avoid "all familiarity, and unnecessary conversation."[10] Be friendly, but not *too* friendly. And it wasn't just the Philadelphia & Reading that advised their workers about being courteous. Harvey & Company, which arranged excursion trips to California, stressed in its brochures that "ladies and children traveling without escort are as well cared for as though accompanied by personal friends."[11] To increase the number of women passengers, railroad advertising worked hard to popularize the notion that women could expect respectful treatment when traveling by train, and if they chose to take the train the companies made sure they would get it.[12]

PHOEBE SNOW

The Delaware, Lackawanna & Western Railroad, which you met in the last chapter and is now known simply as the Lackawanna, had a reputation for being one of the cleanest railroads. It burned anthracite coal as fuel, which meant that its smokestack emissions were less sooty than other railroads that burned other types of fuel, such as other types of coal or wood. The Lackawanna used this fact in its marketing to attract ridership. To tout its cleanliness, some of its advertisements showed an image of Mark Twain (Samuel Clemens) dressed in his perfunctory white duck suit riding along on one of its trains. The ad included a written note from Clemens to a friend saying that he had been traveling on the Lackawanna and that his suit remained as white as when he boarded the train.

More popular and persuasive than the Mark Twain ads were the Lackawanna's marketing campaign that employed a young fictional character named Phoebe Snow.[13] In print ads and posters Miss Snow always was shown in a perfectly white dress. Like Clemens's white suit, it was used to symbolize the railroad's cleanliness. A reasonable subtext is that the white dress also implied her innocence: If the young and innocent Phoebe Snow could travel comfortably and safely on the Lackawanna, so could any woman.

The Phoebe Snow ads combined the theme of the last chapter and this one: Phoebe not only could travel alone and in safety, but in most of the posters the car in which she was riding was shown passing through bucolic landscapes or near some babbling brook. Some ads even showed her engaged in some outdoor activity. In one Phoebe is canoeing in a stream in the Mountain and

Lake Resorts, which was located in rural Pennsylvania and serviced by the Lackawanna. This ad is a twofer: it informs the observer that the Lackawanna can transport you to this refuge of rest and relaxation, and it reinforces the idea that women can enjoy nature's restorative benefits, apparently even if no man is present. If the young Phoebe could enjoy a canoe trip by taking the Lackawanna, why shouldn't other women?

This all may seem sexist, but at the time the point was to appeal to women who, maybe for the first time in their lives, now had the opportunity to escape the protective bubble of their homes and their husband and family's protection. Train travel opened the door to the world and the activities if offers. As the twentieth century began, more and more women were traveling, visiting the outdoors, and taking up recreational fishing. These themes ran through much of the industry's advertising.

THE NEW WOMAN, RAILROADS, AND FISHING

By the end of the 1800s, the act of a woman taking a train trip, sometimes even for great distances, was not considered a breach of social norms. It was the era of the "new woman." Even household products that were used helped define what characteristics defined the new woman. Consider the following ad copy for Pearline, a household laundry soap, that appeared in the popular *Ladies Home Journal* and thus reached a large and diverse female audience.

> What a Difference between the WOMAN who is wedded to old-fashioned ideas and she who is bright enough to appreciate a new one. Everybody is striving to get something to make life easier—often it is right beside them—those who are bright enough to embrace it get the benefits, whose who don't go backwards—their work grows harder. Pearline makes life easier and cleaner. Washing and cleaning done with Pearline has about enough work in it to make it good exercise—but not enough to tire the body or ruffle the temper.[14]

Just by using this product (and similar ads can be found for other products of the day), a woman could define how progressive and modern she was. Relate this to the railroads' promotions aimed at making women feel less apprehensive—not just about safety, but also social acceptability—about train travel. Mix these ideas with the emerging notion that it was socially acceptable, even recommended, for women to participate in outdoor sports, especially the "non-violent" types such as fishing and bicycling, and a movement begins to take shape. "Angling represented a respectable sport for Victorian women," writes historian Jenn Corrine Brown, "as long as they maintained proper gender boundaries."[15]

Breaking such boundaries clearly was the topic of the cheeky cover illustration for the 25 October 1890 issue of *Harper's Bazaar* that shows a young woman, fly rod in hand, holding aloft a stringer not of fish but of men. The C. S. Reinhart illustration is entitled "A Successful Catch."[16] The times were changing. Fishing, especially fly-fishing, was considered the "gentle art" and a sport suitable for women. Many from various walks of life were drawn to it. With railroads actively seeking to attract female riders, and with the rising appeal of sport fishing in general, it is logical that more women would take up the sport and begin visiting some of the country's noteworthy fishing locales.

Editors of popular outdoor magazines took notice of this trend and responded by including fishing articles that featured women or were written by them. "While angling had developed into a respectable female pastime," observes one angling historian, "it too, benefited from the impetus given to the inclusion of women in sporting magazines. Fisherwomen now had a vehicle through which they could publicly share their experiences and knowledge of the sport to not only a new generation of women, but also to a wider circulation of supportive and appreciative men."[17] *Rod and Gun* published a series of articles by Anna Williams recounting her fishing trips.[18] An article in *Forest and Stream* by Mrs. F. Cauthorn informed readers that, given the opportunity, "a woman will soon learn to love the ripple or roar of a trout stream and the sound of the reel as much as her husband or big brother, and look forward expectantly to a 'day off' in the mountains enticing the wary trout."[19] In another *Forest and Stream* article, after posing the question "Why should it be considered outré for ladies to fish?," Lucy Tomlin of Minnesota describes her ability to catch trout and bass as effectively as her husband.[20] *Forest and Stream*, a staunch promoter of the emerging fishing ethic, even began publishing a regular column devoted to issues about women in the outdoors.

But let's not lose sight of how this all came about. Women would not have experienced these fishing adventures were it not for the fact that in nearly every instance it was necessary for them to take the train to reach the stream or lake that they wrote about. Railroads in a very direct way encouraged this trend through their advertising campaigns, which not only showed women riding the train but also women in fishing locations made accessible by taking the train.

As more women found their way into rivers and streams in search of trout or other game fish, some women emerged as experts in various aspects of the sport. Mary Orvis Marbury, daughter of Charles F. Orvis, the founder of the Orvis Company, was well known for her knowledge of fly tying and fishing. She published articles dealing with various aspects of fly-fishing in the leading outdoors publications of the day, including *American Angler, Forest and Stream*, and *Outing Magazine*.

The writer Mary Trowbridge Townsend chronicled her fishing adventures in *Outing Magazine*, a national publication that covered a myriad of sports. In one of these articles, Townsend writes about fly-fishing for trout in Yellowstone Park.[21] Her article chronicles her time in Yellowstone and made an impact. "People had been sport fishing in the park ever since its founding," notes historian Paul Schullery. But, he continues, "most of those earlier accounts were published too obscurely to stick in the public mind the way her account seems to have."[22] In a later article Townsend recounts her experiences of fishing for brook trout in eastern U.S. and Canadian streams. In both articles she reveals her knowledge of the sport, but more importantly shows that a woman can be as adventuresome as any man when it comes to plying her fly-fishing skills and enjoying the beauty of the rugged outdoors. Of course, in both cases she could only have accomplished these experiences by first traveling by train.

Articles like these—and there are others—not only introduced women to fly-fishing but helped allay any social stigma that may still have been attached to women taking part in outdoor sports. These articles lowered the barriers of entry for the average woman to take part in the once male-dominated sport of fishing. As these barriers fell, more women began fishing, many visiting Yellowstone Park, others heading for trout streams in Colorado or Maine or Michigan. To be sure, not everyone wanted to fly-fish or could afford to make the trip to some fairly remote trout haven. But the articles by women and about women fishing gave rise to women's participation in sport fishing, whether it was for trout or some other species in the local lake. And while railroads may not have been the only reason that explains the increasing number of women fishing, surely they must get substantial credit for encouraging the emerging trend.

One historian has suggested that an "area of influence and independence that angling provided to Victorian women was through the public role of promoter and marketer."[23] Cornelia Crosby of Maine fit that description to a tee. A nationally recognized outdoorswoman and professional guide, Crosby wrote extensively about hunting and fishing. Early in her career she was given the nickname "Fly Rod" by one of her editors because of her prowess with the fly rod. The moniker stuck, and her column "Fly Rod's Notebook" appeared in major newspapers in New York, Boston, and Chicago, and in national sporting journals such as *Shooting and Fishing* and *Forest and Stream*.

Crosby is important to this story because during her career she worked as a spokeswoman for the Maine Central Railroad.[24] Briefly, here is how the two came together. Crosby thought Maine should have some representation at the First Annual Sportsman's Exposition, to be held at New York's Madison Square Garden in 1895. Unable to fund an exhibit herself, she approached Payson Tucker, the president of the Maine Central Railroad. Crosby's pitch was to have a Maine exhibit that consisted of a log cabin replete with what anglers and hunters could expect to find in Maine: mounted fish, deer, and moose. Several Maine guides would be on hand to answer questions and provide promotional material to interested attendees. Crosby also would be there promoting Maine and, one assumes, tending to her own business interests. The bottom line was that it would only happen if the Maine Central Railroad paid for it.

Tucker was so taken by Crosby's proposal and her energy that he agreed that his railroad would cover all expenses for the exhibit. He also put Crosby in charge. But the agreement came with one proviso: Crosby would promote the Rangeleys, Moosehead, the West Branch of the Penobscot, Grand Lakes, and other prime Maine destinations for fishing and hunting, but she was *not* to mention his railroad. Anyone visiting the exhibit would see no evidence that the Maine Central was its sponsor. Why wouldn't the president of the railroad paying for an exhibit in such a major exposition want its name front and center? Tucker was shrewd enough to know that any tourist business generated from the exhibit would need to ride on the Maine Central to reach any one of these destinations.

By all accounts the exhibit and Crosby's management of it was a great success. The surge in the number of visitors, tourists, hunters, and anglers to Maine following the exposition was notable. One estimate is that the exhibit generated an extra five thousand visitors in one summer alone. Of course, this meant an increase in the number of passengers for the Maine Central. Given this successful partnership Tucker hired Crosby to a position in the railroad's publicity department where she worked for a number of years, promoting Maine, increasing the ridership of the Maine Central, and attracting more women to sport fishing.

Another Maine railroad took the idea of cross-promotion to a different level. Beginning in 1901 the Bangor & Aroostook Railroad began publishing a magazine called *In the Maine Woods*. An annual publication that ran about two hundred pages, the magazine included the maps and schedules for the Bangor & Aroostook as well as ads for everything someone heading to the Maine woods might need: guides, hotels, taxidermists, sellers of guns, fishing equipment, camping equipment, and more. Because the locales for the various activities included in the articles were reached by taking a Bangor & Aroostook train, readers naturally would link Maine's outdoor attractions to the railroad.

This is relevant in this chapter because the magazine often included articles by and about women involved in the outdoors. One such article by Mrs. James A. Cruikshank was titled "The Woman's Standpoint." In it Mrs. Cruikshank states that

> as a rule, modern women are supposed to prefer ease rather than exercise, conventionality rather than originality and luxury rather than pioneer simplicity. Yet the woman who has camped in Maine, or who has made one of its many wonderful canoe tours, living the simple woods life, wearing old clothes, sleeping on balsam boughs with the sky for roof, photographing wild creatures, fishing for salmon or trout, or hunting big game, comes home and tells her story with all the enthusiasm of the school girl.[25]

Articles like this were important in convincing more women to take part in outdoor activities like fishing. Sometimes the cover of *In the Maine Woods* featured a woman engaged in one of Maine's outdoor activities. One cover in particular shows a young woman fly fisher who has caught her trout and appears quite adept in landing it by herself. The implication is clear: a female angler can enjoy her time in the stream thanks to the Bangor & Aroostook Railroad for getting her to this fishing paradise.

ADVERTISING TO WOMEN

Fly-fishing literature (and fishing in general) began to recognize the attention that women were giving the sport and so accommodated this curiosity by including articles by and about women fishing. More nudges came from railroads' and, as you will see, other companies' promotional materials. You've already seen some of this in the last chapter—for example, railroad guidebooks and travel brochures extolling the fishing excursions that men *and* women could take. Many railroads' promotional materials showed a woman fly-fishing in a stream, suggesting that women could enjoy this pastime if only they took the railroad's line to an undoubtedly trout-filled stream in some idyllic setting.[26]

Although they are outside of my self-imposed 1900 cutoff, railroad ads from the early twentieth century provide some of the best examples of how the industry combined advertising to entice ridership and the allure of fly-fishing in some delightful setting. I'll use just a few examples.

An ad in the 4 April 1925 issue of the *Literary Digest* for the Great Northern Railway is a wonderful example of how railroads combined themes.[27] The *Digest* proclaimed itself to be the magazine for the "alert woman," which it defined as the modern-day woman who cared for the home but still found time for community work—and, it appears, for recreational outlets. The ad shows a young woman all set to try her hand at trout fishing in a stream in Glacier National Park, which was serviced by the Great Northern. Success in catching trout, the copy claims, is almost assured. Everyone—the experienced and the neophyte alike—will be rewarded with "a beautiful catch nearly any time of day you make the try." But first, of course, you must hop aboard a Great Northern train.

The use of an image of a woman fishing to promote a destination was in a Rock Island railroad ad that appeared in *Munsey's Magazine*, another popular magazine with national readership.[28] The image shows a young couple fishing in an unnamed stream somewhere in the mountains of Colorado. The man in the image is reeling his catch and the woman, rod in hand, is about to take her turn. The lovely image is suggestive enough, but the text provides the hook. It reads like a letter home from the woman to a female friend, sort of a "Wish you were here" note. The text extols the wonders of the Colorado experience, including the excellent fishing. In effect the woman's note is an invitation to visit Colorado and go fishing, which, of course, is best done by booking a trip on the Rock Island's Rocky Mountain Limited.

Lastly, I mentioned earlier how the Pullman Company often emphasized women in their ad campaigns to promote rail travel. A 1936 poster for the company by William Welsh is one of the best examples of using imagery and minimal text to convey the service provided by the company. Since Pullman was not a railroad, the tie-in is to take any railroad that offers Pullman cars.[29] The poster shows a young blonde woman in hip boots, fly rod in hand, playing a trout in waters below a tumbling cascade. Across the bottom appear a series of jumping rainbow trout. The text is sublime: "Vacation Lands Are Calling . . . Go in Safety & Comfort." PULLMAN is prominently displayed in the lower left-hand corner of the poster. This image and limited text say it all: the woman angler is enjoying her time fly-fishing. Whether she is alone or not is up to the viewer. How did she get to this sanctuary of beauty and fish-filled waters? By riding in a Pullman car attached to some railroad company's train. Regardless of the railroad, we know that she traveled in safety and comfort because she rode in a Pullman.

As informative and beautiful as the Welsh poster is, it and the other promotional materials highlighted here—and there are many more—all conveyed the same message: women should travel by train. The not-so-subliminal message was that they also should consider and would probably enjoy sport fishing, especially fly-fishing, as a diversion. The fact that in some of these ads the women appear to be traveling alone further builds upon the notion that the adventuresome new woman of the times need not be accompanied by a male escort to experience travel and the outdoors.

A REALITY CHECK

It is useful to recognize that the foregoing discussion of women traveling and taking up fishing reveals to some extent class privilege. During the latter half of the 1800s, as I have noted elsewhere, there was a growing number of families that today we would classify as middle income. Along with this increase in income often came more leisure time, time that could be devoted to recreational activities such as fishing and traveling. Throughout much of this period (and, arguably, well into the twentieth century), women were the major providers of home care. Even in the middle-class family, this often gave the man more time to engage in leisure activities but still constrained the woman from engaging in these sports equally. In that sense women subsidized men's leisure.[30]

I bring this up because the liberated Victorian woman depicted in railroad ads and targeted by railroads as potential passengers headed for Yellowstone or other exotic destinations most likely were of a socioeconomic class that could afford to take extended holidays. Yes, there certainly were more women in all income strata taking up fishing as the nineteenth century rolled into the twentieth. And there was a small subset of women who became guides (like Cornelia Crosby) and who chose to wear attire more efficient (i.e., men's clothes) for fishing. But, for the most part, the majority of women who were actively traveling to visit fishing locales and engage in the gentle art of fly-fishing did not come from the lower echelons of the income scale.[31]

The Fishing Industry Responds

Companies in the sport fishing business responded to the increased number of women taking up the sport. Beginning in the late 1800s and into the next century, fishing equipment designed specifically for women began appearing in the catalogs of the most famous tackle companies. Look through Melner and Kessler's *Great Fishing Tackle Catalogs of the Golden Ages* and you see companies like the Orvis Company of Manchester, Vermont, offering a "ladies' fly rod" in their 1884 catalog. The rod was "designed for the use of Ladies . . . fitted with CORK HAND-PIECES that do not require as much strength to hold the rod firmly."[32] By the early 1900s not only were catalogs including fly-fishing equipment designed for women, but also women began appearing in the ads. Heretofore, only males were shown, since, after all, it was a male pastime. In one Abercrombie and Fitch catalog a woman, rod in hand, is modeling the "Nepigon," a Norfolk-style fishing coat complete with Knickerbockers. Unlike her nearby male model, however, she does not have on waders. Similarly, in a catalog for the New York Sporting Goods Company a man is shown in hunting attire, and a woman is decked out for fishing, though, again, she is wearing a skirt and not waders.[33]

These ads illustrated a problem that many women anglers faced: How to wear a dress and waders at the same time? Most waders made it impossible to stuff one's dress into them. The "fashion" for women anglers, who apparently could not buck societal norms enough to wear pants, was a skirt that looked as if it was tucked up to the knees. This is the style that one sees in, for example, the illustrations that go with the articles by Townsend discussed earlier. I am sure it was not universal, but I have not seen any photograph or ad from this era where a woman is actually wading midstream. Over time this would change, as women simply began to wear pants when fishing so that a pair of waders would fit. Whether it was advertising fly rods or fishing attire, women were becoming a large enough segment of the market to call for special attention.

Women and the Conservationist Movement

I want to build on the idea presented in the previous chapter that railroads were more influential in advancing the conservation movement than many modern readers might think. Although the clarion call for conservation had begun in the mid-1800s, the movement gained traction in the latter part of the century. You have already seen that in order to promote the scenic views along their routes railroads took it upon themselves to protect it: billboards were taken down, rights-of-way provided a natural environment, and depots were beautified through the railroad garden movement. Railroads also supported the development of the national parks and the preservation of land.

By lowering the barriers to fishing (e.g., easier access, cheaper travel costs), railroads helped grow the ranks of anglers, men and women. As the number of recreational anglers swelled, so, too, did the subset of those individuals who became concerned about the condition of the nation's fishery. The prevalent practice of overfishing or leaving caught fish on the banks to rot damaged the fish population in many locales. To preserve the stock of fish, and therefore the ability to fish and successfully catch them, more and more anglers began to recognize that preserving and protecting the object of their pastime was important.[34] Conservation historian John Reiger has argued that the increasing number of fishing enthusiasts "provided [many with] that crucial first contact with the natural world that spawned a commitment to its perpetuation."[35] If the actions of railroads gave rise to the number of anglers, then by extension railroads were important in nurturing the conservation movement as it regards sport fishing.

By growing the ranks of women engaged in fishing and other outdoor activities, railroads may have even given the conservation movement a bigger boost. We know, for example, that the newly created national parks in the West were a popular draw for women. Glenda Riley points out that the number of women visitors to the Yellowstone National Park began to increase notably in the 1890s. This came about once "railroad companies, hotels, chambers of commerce and the U.S. Park Service combined their efforts" to promote national parks through the use of "a variety of brochures, travel posters, free maps, and works of art [which] lured tourists to the parks."[36] Why is the rising number of women visitors significant? First, it showed that societal constraints on women traveling by train were wearing away. Second, when many of these women returned home, they actively encouraged other women to make the same journey. This often was accomplished through local women's clubs, many of which were established with specific topics in mind, such as increased environmental degradation and the loss of fish and wildlife. Many women and women's clubs grew to become outspoken advocates for conservation efforts.

These women's clubs became more organized, even nationally. This is exemplified by the 1890 establishment of the General Federation of Women's Clubs. With increased participation, their voice on many causes, including conservation issues, grew louder. As the 1800s rolled into the Progressive Movement, many women's clubs became actively engaged in the conservation movement, even though their influence was more often plied behind the scenes.[37] Riley's analysis indicates that "environmentalism would have been far less effective had it not been for the thousands of women who supported it."[38]

How is this related to railroads? This expansion of women's perspective was largely accomplished because railroads made their travel to these once unattainable locations possible. Once women could and did journey to Yellowstone or Glacier or the wilds of Maine and Michigan, they recast "the public perception of 'wilderness,' [making] both the outdoors and concern about its future palatable to a wide audience."[39] Indeed, one travel pamphlet boasted that "thousands of women annually visit Yosemite National Park and enjoy Yosemite Tours without escort."[40] Women traveling to Yosemite and other natural wonders in the safety and comfort of the train were being made aware of the mounting environmental problems that continued use and sometimes abuse of the country's natural resources was having. This awareness prompted many to advocate for a variety of conservationist policies, some of which focused on protecting the fish population.

A Final Note

By the late 1800s, women were taking up the sport of fishing in ever-increasing numbers. Some were making significant contributions to the sport. Cornelia "Fly Rod" Crosby gained fame for her writing of fishing and hunting in Maine. Sara Jane McBride won widespread attention for her "Metaphysics of Fly Fishing," which appeared in *Forest and Stream* in 1876. And, of course, Mary Orvis Marbury would gain fame for, among other accomplishments, her massive tome *Favorite Flies and Their Histories* (1892). These women were pioneers in the field of fly-fishing, providing acceptance for women in general to take part in this sport.

For the New Woman breaking free from Victorian-era constraints, more freedom to travel meant a new independence. This was expressed in increased travel for tourism, and to engage in pursuits like sport fishing. Railroads promoted innovations in travel (e.g., sleeper cars, dining cars) that encouraged more women to travel by train. Making travel easier lowered a significant barrier to women's participation in sport fishing: being able to access fishing locales, even those found far from home. The railroads' targeted marketing campaigns sent a clear message: women could travel in safety and comfort when they stepped onto a train car, no matter their destination.

Railroads enriched the diversity of those involved in sport fishing by breaking down gender barriers that had prevented women from traveling into the great outdoors. Their promotional campaigns not only encouraged women to travel and to try fishing but also increased the number of anglers and strengthened the emerging conservation movement.

ONE FINAL NOTE

Who would have thought that the convergence of two technologies—fish culture and rail transportation—in the second half of the 1800s would forever affect both fish management and sport fishing in the United States. Railroads gave the fledgling U.S. Fish Commission the means to experiment with transplanting different species of fish throughout the country. Not only were there a wide variety of species moved here and there, but millions of them. Some of these experiments were successful, if success is measured in terms of a species adapting to its new surroundings and flourishing. Today rainbow trout are seemingly everywhere. Carp are also ubiquitous, but you can decide if that represents a success or a failure. Other attempts to transplant fish were spectacular failures. Pacific salmon sent by rail to the East did not take hold in eastern rivers, nor are they found in the river systems of the Midwest, where the Commission tried to introduce them.

For the modern reader, it is perhaps easy to scoff at these brazen interventions into natural processes. But put yourself into a world in which the stock of fish, especially along the coasts and in some freshwater locations like the Great Lakes, was in jeopardy. Fishing regulations to stem the overharvest were unpopular and seldom enforceable. If the public, especially commercial fishermen, could not regulate its harvest of fish, the Commission's work was considered laudable: use fish culture to augment nature's efforts at producing enough fish to feed a growing population. What history tells us is that, though well-intentioned, the hubris of science can lead to unintended and sometimes harmful consequences. Even though Commission officials fairly quickly realized that the widespread introduction of some species was misguided, it became impossible to undo the sometimes devastating and long-lasting effects on habitat and on native species. Just because one can transport and stock different species of fish across the country does not mean that one should.

Through it all the evidence overwhelmingly shows that this inspired and publicly popular endeavor would not have been possible were it not for the assistance of many railroad companies. Hundreds of railroads transported the Commission's cargo of fish cans and their attendant fish messengers thousands of miles every year. In the latter half of the 1800s, railroads also were moving the Commission's (and state commissions') fish cars tens of thousands of miles a year. What is astonishing is the fact that these were private companies providing such aid to the government at significantly reduced rates or, more commonly, for free.

The joint endeavor by the Commission and the railroads did not solve the problem they set out to alleviate. But their actions had another unexpected benefit: making sport fishing a popular and enduring pastime. By introducing various species of fish into new waters, the actions of the Commission and the railroads were a boon to anglers. Anglers across the country could now fish for species that were beyond the reach of the average individual. Instead of reading about catching fish like brook or rainbow trout, anglers in many states could ply their skills against these exotic members of the "finny tribe." This increased accessibility came not only because the Commission was planting rainbow trout wherever they could, but also because railroads gave anglers the opportunity to visit locales where these and other sought-after species lived. With long-distance travel now routine, anglers from around the country could reach desired fishing destinations from one side of the country to the other. Making the fishing experience possible to an ever-widening segment of the population was instrumental in growing the general interest in sport fishing.

Because of the railroads' involvement, sport fishing evolved from a pastime of the well-to-do to a recreational activity enjoyed by the general public. Besides its work with the Commission, railroads acted in other ways to advance the popularity of sport fishing. Railroads understood that the scenery along their routes was an asset to their transportation business: a pleasing view as the train rolled along added to the passenger's enjoyment of the journey. Consequently, railroads not only advertised the attractiveness of the scenery, but acted as stewards of the landscape through which their tracks ran. Providing passengers with a picturesque ride was good for business. It also raised awareness among passengers of nature's beauty and, therefore, the reason why it should be protected. It may have been tangential to their profit motive, but railroad advertising and behavior motivated individuals, especially those living in urban areas, to visit the outdoors and engage in various activities, like fishing. Of course, once the public's enthusiasm for fishing took hold, it soon became clear that the railroad was the best mode of transportation by which to visit those desired fishing destinations.

In order to boost ridership, railroads advertised their ability to get individuals—especially those in urban areas—to a growing list of rural destinations. From its earliest days, the railroad was considered to be unequaled in getting weary urbanites into the restorative arms of nature. Its extensive advertising and marketing campaigns promoted the outdoors as a destination and fishing as a recreational pastime enjoyed by all; men and women, novice and expert. Because of their vast network of tracks, railroads became the only sensible means by which the interested angler could access fishing opportunities once considered too remote or too distant. Railroads opened the nation for the fishing enthusiast with fast and low-cost travel.

Railroad advertising was aimed to increase ridership, and much of it was targeted at women. As the confining rules of Victorian society began to fade, railroads worked hard to establish an environment in which women could experience the joy of rail travel. Railroads facilitated this societal shift in various ways. One was offering special cars in which only women (or women with their children) could ride. Railroads also focused their advertising to ease any anxiety a woman might have about taking the train, even unaccompanied by a male escort. Claims made in advertising were backed with action. Railroad employees often were required to assist women, including dealing with unwanted advances from male passengers. Indeed, by the end of the 1800s, women, often unaccompanied by a male companion, were traveling by train in far greater numbers than ever before.

Another change in public opinion that railroads endorsed was the popular belief that women could and should take part in outdoor activities. One of those considered "acceptable" was fishing. In response to this trend, railroads' promotional materials often showed women fishing in some pristine stream. The purpose of the ads and the service that backed them was to increase interest in travel to popular areas like Colorado or Maine, where fishing was a popular tourist activity. With fishing condoned as a suitable pastime for women, railroads strove to make sport fishing more popular among women. By getting more women to explore the outdoors and the recreational opportunities it offers, railroads also expanded the base of the conservation movement that would emerge with greater force in the 1900s.

My story ends mostly in 1900. From that point on, the fate of the two central players would take very different arcs. The railroad's supremacy in transporting people lasted only a few more decades. The number of passengers and passenger miles peaked in 1920 and, except during World War II, steadily declined after that. Another new technology, the automobile, would cut deeply into the railroad's near-monopoly over moving people around the country. Railroads remained as the major player in moving freight, though trucks have steadily eaten into this market over time. Of course, in more recent years an even newer transportation technology, the airplane, has swallowed up its share of the passenger and the freight markets.

What of the Fish Commission? The Commission's administrative location in government shifted over time. It became part of the Commerce Department in 1903, was transferred to the Department of Interior in 1939, and in 1940 merged with the Bureau of Biological Survey to become what today we know as the U.S. Fish and Wildlife Service. Along with administrative changes came an evolution in its portfolio of work. As one observer sums up the Commission's history, "the first one hundred years of [the Commission's] fisheries conservation were a carrying out of Spencer Baird's legacy of sciences, stockings, and sportsmen, and the next fifty reflected [Rachel] Carson's ethos of ecology, imperiled species, and loss of habitat."[1]

Though history altered the paths of these two institutions, the inescapable fact is that, without their cooperative actions and the railroads' own campaigns to buoy interest in travel and the outdoors in the 1800s, sport fishing in the United States would look much different than it does today.

ACKNOWLEDGMENTS

I have benefited from the input of numerous people during the course of completing this project. Thanks to Dan Brennan, Jenn Corrine Brown, Anders Halverson, Lynn Morrow, Mike Kruse, Bruce Petersen, John Reiger, Paul Schullery, Craig Springer, and two anonymous reviewers for their comments and suggestions. I may not have succeeded in answering all the questions raised or carried out every suggestion made, but making me reexamine my arguments made this a better product. I also thank Professor Tim Schoenecker, who, though I'm sure he'll disagree, made this book possible. I, of course, retain an author's responsibility for any errors or omissions.

Thanks to those at the University of Nebraska Press who improved an earlier product. Emily Shelton made everything read better. Abbey Frankforter shepherded the manuscript through the production process. And special thanks to my editor, Clark Whitehorn, who thought the topic interesting enough to guide me from rough manuscript to this finished product.

I have addressed some of the topics in this book in earlier articles. Several appeared in the *American Fly Fisher*, published by the American Museum of Fly Fishing in Manchester, Vermont. My thanks to the journal's editor, Kathleen Achor, for greatly improving earlier versions of those articles and seeing them through the publishing process. I thank the Museum for permission to use some of the material found in those articles. I also would like to thank Dan Cupper, the editor of *Railroad History*, who made me appreciate just how complex U.S. railroad history really is.

Gail Heyne Hafer has endured years of listening to me prattle on about some latest find in the history of fish culture or railroads. Even so, she was kind enough to read and comment on drafts of these chapters. My sincerest thanks seem hardly sufficient.

NOTES

1. A Brief History of Railroads

1. For more on the history and impact of the Erie Canal, see, among others, Atack and Passell, *New Economic View of American History*. For an overview of public works in America, see Goodrich, *Government Promotion of American Canals and Railroads*.
2. This is the assessment, a somewhat biased one, of Poor, "Sketch of the Rise," in *Manual of Railroads*, 13. Poor published the *Manual*, an important publication about the railroad industry, listing companies and statistics about them, in addition to data on the financial conditions, bond issues, and the like, of the federal government and states.
3. Stover, *American Railroads*, 15.
4. Caldwell, "On the Moral," 293. Besides being an influential advocate for railroads, Caldwell is, perhaps unfortunately, remembered as someone who tried to establish phrenology—a study of the shape and size of the cranium to determine one's mental ability—as part of the medical profession in the United States.
5. Caldwell, "On the Moral," 296.
6. Reported in "St. Louis Railway Commenced," 142. Such imagery was not unique to railroads. Representative John C. Calhoun of South Carolina exhorted Congress to increase funding for new wagon routes with the words "Let us bind the Republic together with a perfect system of roads and canals." Quoted in U.S. Federal Highway Administration, *America's Highways*, 19.
7. Minor, "Railroad Journal," 1.
8. Webster, "Opening of the Northern Railroad."
9. Clifford F. Thies offers an interesting take on the interplay of private and public incentives railroads in his "American Railroad Network During the Early 19th Century," 229–61.
10. Baker, *Formation of the New England Railroad System*, 3.

11. Ward, *Railroads and the Character*, 75. By the mid-1880s the governments of New York and Pennsylvania had accrued debts related to underwriting railroads in their states amounting to $8,000,000 and $38,000,000, respectively. For some perspective the modern dollar equivalent for New York is roughly $265 million and for Pennsylvania about $1.3 billion.
12. Poor, "Sketch of the Rise, Progress, Cost, Earnings, Etc.," in *Manual of the Railroads*, 15.
13. For a discussion of the surrounding events, see Wallis, "What Caused the Crisis of 1839?"
14. Taylor and Neu, *American Railroad Network*, 8–9.
15. Douglas's political career included such accomplishments as leading the passage of the Compromise of 1850, introducing the Kansas-Nebraska Act in 1854, and, most notably, losing to Abraham Lincoln in the 1860 presidential election. King, who had earlier served as a U.S. representative from North Carolina, would later serve as vice president under President Pierce.
16. Stover, *American Railroads*, 45.
17. My focus is on the railroad, but another technological advance that needs mention is the telegraph. With telegraph poles following rail lines and with posts in train stations, the telegraph revolutionized the market for information. Far-flung markets now were connected by the railroads and by the telegraph. For a history see, among others, Blondheim, *News over the Wires*.
18. An excellent source is Ambrose, *Nothing Like It in the World*.
19. Stover, *Iron Road to the West*, 232.
20. By most accounts New York City merchant Asa Whitney was the first to present a serious proposal for a railroad to the Pacific in the mid-1840s. Whitney's plan proposed a railroad line that stretched from the western shore of Lake Michigan to the Pacific Ocean. The idea was to reduce shipping time (and cost) between New York and the West Coast, and thus China. See Brown, "Asa Whitney," 209–24.
21. Details about the Act of 1862 can be found in, among others, Ames, *Pioneering the Union Pacific*, especially chapter 1.
22. This route is roughly similar to the one that had been adopted at the national railroad convention held in St. Louis in the fall of 1849. See Cotterill, "National Railroad Convention," 203–15.
23. These funding provisions are detailed in Sabin, *Building the Pacific Railway*.
24. Ames, *Pioneering the Union Pacific*, 14.
25. An informative review of the founder's backgrounds and the early days of the Central Pacific is presented in Sabin, *Building the Pacific Railway*. See also Lewis, *Big Four*.

26. Sabin, *Building the Pacific Railway*, 31.
27. Stover, *American Railroads*, 69.
28. The federal land grants given to the Union Pacific blatantly ignored any property rights held by Native American tribes to the land through which the tracks would run. In the eyes of the government, the tribes lacked any legal claim to the land, and thus it was considered the property of the U.S. government. Consequently, the government had the legal right to dispose of it as it saw fit.
29. Depending on the state of luxury in which one wished to travel, a trip from New York to Sacramento became accessible to a large segment of the population. The cheapest (and least comfortable) fare was the $65 "emigrant" rate. Stepping up to second class raised the fare to $109, and a first–class ticket was $130. Passengers could pay extra for a sleeping car—$2 a day and $2 a night—and meals at dining stations were another dollar. Children under five often traveled for free. For some perspective, in today's dollars the fare would run from about $1,900 for the emigrant rate to $3,800 for the first-class passenger. The original fares and the passenger figures are from Sabin, *Building the Pacific Railway*, 308.
30. Prior to the transcontinental railway, the Overland Stage Express Company carried about a thousand pounds of mail every day from Omaha to California. Without any incidents the trip took sixteen days to complete, at least during the eight months out of the year that it was available. For this service the government paid Overland a little less than $2 million per year. Once the transcontinental line opened, however, the railroad delivered six thousand pounds of mail at an annual cost of $513,000. The opening of the transcontinental railroad had another important effect. Now the government could increase protection for westbound settlers by dispersing men and materiel to Western military outposts much faster than ever before. These are from Sabin, *Building the Pacific Railway*, 307–8.
31. R. L. Duffus, "Some Political Scandals in the History of the Nation," *New York Times*, February 10, 1924. There are many sources for this discussion. See, among others, White, *Railroaded*. A contemporary assessment is provided by Crawford, *Credit Mobilier of America*.
32. "The King of Frauds: How the Credit Mobilier Bought Its Way Through Congress," *Sun*, September 4, 1872, 1.
33. Railroads and scandalous behavior were not strange bedfellows. Significant owners and investors in railroads, such as Cornelius Vanderbilt, Jay Gould and Jim Fisk, and J. P. Morgan, among others, often used their interests in railroad companies to manipulate financial markets in their favor. Sometimes the confluence between the railroad industry and the financial markets led to serious

national complications. For example, Jay Cooke & Company invested heavily in the building of the Northern Pacific Railroad. When higher than expected construction costs and weak revenues meant that the company could not service its financial obligations, the Northern went bankrupt. So did Cooke's financial company. The failure of the railroad and Cooke's company created a ripple effect through the financial market that helped trigger the financial panic of 1873 and the economic downturn that lasted for years.

34. One example is DeBow, "Additional Remarks by the Editor," 485–95. DeBow was the influential editor of the *Commercial Review*, a widely read Southern publication that covered all facets of Southern life, including its unapologetic support of slavery.
35. Stover, *American Railroads*, 120.
36. Stover, *American Railroads*, 261.
37. Even "when government stock ownership entitled states and localities to representation on boards of directors and to vote in stockholders' meetings, such governments generally adopted a passive role, deferring to private [railroad] leadership." Trelease, "Passive Voice," 174.
38. Poor, *Manual of the Railroads of the United States for 1872–73*, 31.
39. Dartnell, *Report on the Gauge*, 20–21. Cited in Taylor and Neu, *American Railroad Network*, 50–51.
40. It is worth noting that most of the railroads in the South often used a smaller gauge. In 1863 the South had little to say about what got passed in the two houses of U.S. government. This is not the first or the last incidence of ignoring Southern interests. Indeed, the route of the transcontinental railway reflects a decidedly pro-Northern bias. For example, even though by the start of the Civil War railroads "served quite adequately all the states east of the Mississippi and few areas of substantial population were much removed from the sound of the locomotive whistle," by the late 1850s only three connections between railroads north of the Ohio River and the South existed. Stover, *Iron Road to the West*, 232.
41. Taylor and Neu, *American Railroad Network*, 82. Even though there was general industry acceptance of this standard gauge, it wasn't until October 1896 when the Committee on Standard Wheel and Track Gauges of the American Railway Association recommended that a gauge of 4 feet 8 ½ inches should be the industry standard.
42. Everything you might want to know about the development of the passenger car can be found in White, *American Railroad Passenger Car*. The following discussion is based on those volumes.

43. Barky, "Invention of Railroad Time," 12–22.
44. Vernon, *Travelers' Official Rail Way Guide*, 1868, iii.

2. A Brief History of Fish Culture

1. Roosevelt, *Fish Culture*, 14.
2. Shanks, "Fish-Culture in America," 721.
3. Bertram, *Harvest of the Sea*, 474.
4. Whelan et al., "Tracking Fisheries Through Time," 393–94.
5. U.S. Bureau of the Census, *Statistics of the United States in 1860*.
6. Attributed to Gilbert Fowler, cited in Waldman, *Running Silver*, 99.
7. U.S. Bureau of the Census, *Statistics of the United States in 1860*, 534.
8. Historian John Reiger notes that a relative of Marsh, George Franklin Edmunds, then the speaker of the Vermont Legislature, was influential in getting the Vermont legislature to commission Marsh to write his report. Edmunds was an avid angler and deeply concerned about conservation issues, such as the depletion of the fish stock. Edmunds, later as a member of the U.S. Senate, helped create the U.S. Fish Commission and wielded enough influence to see that his friend Spencer Fullerton Baird would head the newly minted commission. Edmund's name will appear later in our treatment of the commission's creation. See Reiger, *American Sportsmen*, 80.
9. Marsh, *Report*, 11. Marsh's report did not focus only on the fishery, but also analyzed the effects on wildlife in general.
10. Webster's comments about how one should assess the potential damages caused by trains offer such an example.
11. Marsh, *Report*, 11.
12. Marsh, *Report*, 9.
13. Roosevelt, *Superior Fishing*, 185–86.
14. Marsh, *Report*, 12–13.
15. Bertram, *Harvest of the Sea*, 474.
16. Quoted in Whelan et al., "Tracking Fisheries Through Time," 405.
17. Jerome, *Third Report of the Superintendent*, 53.
18. Marsh, *Report*, 13. Shanks argues that "the present great scarcity of salmon, trout, and shad, the finest of fish in the estimation of both sportsman and gourmand, owes almost entirely to the increase of our industrial enterprises." Shanks, "Fish-Culture in America," 722.
19. Waldman, *Running Silver*, 99.
20. Whelan et al., "Tracking Fisheries Through Time," 405.

21. Marsh, *Report*, 15.
22. Hallock, *Vacation Rambles*, 29; quoted in Burd, "Imagining a Pure Michigan Landscape," 36.
23. J. H. Henshall, "Black Bass in Trout Waters," *Forest and Stream*, July 23, 1886; quoted in Halverson, *Entirely Synthetic Fish*, 14.
24. Marsh, *Report*, 16.
25. Schullery, *American Fly Fishing*, 125.
26. *Forest and Stream*, 5 May 1892, 425.
27. A Mr. Shaw of Drumlaning, Scotland, deserves mention here. It appears that in 1837 Mr. Shaw took spawning salmon from the river Nith, harvested eggs from the females, and impregnated them with milt from the males. He then raised these artificially produced salmon to the age of two, whence they were released back into the river. Shaw's work never registered because his was more of a scientific experiment than a means to raise salmon for food or to replenish the stock of salmon. See Shanks, "Fish-Culture in America."
28. Remy is given credit as the inventor of pisciculture in France. The term would be translated to "fish culture" in English-speaking circles, a change that raised the level of competition and animosity between international rivals in the field. The development of fish culture would expose a deep vein of nationalism as experimenters in different countries adopted and then claimed as their own invention Remy's process, as Remy had done with Jacobi's findings. For a fascinating analysis of the multifaceted nature of the evolution of fish culture, see Kinsey, "'Seeding the Water,'" 527–66.
29. Kinsey, "'Seeding the Water,'" 536.
30. Kinsey, "'Seeding the Water,'" 538.
31. Garlick, *Treatise on the Propagation*, 5.
32. The following draws on Black, "Seth Green," 1–24; and Hedgepeth, "Founders of Fish Culture," 11–14.
33. Stone, "Some Brief Reminiscences," 337.
34. Stone, "Some Brief Reminiscences," 337.
35. "The Father of Fish Culture: Seth Green's Ideas about the Finny Tribe and Some of his Varied Experiments," *Clinton Advocate* (October 11, 1883), 3.
36. Stone, "Some Brief Reminiscences," 338.
37. Green, *Trout Culture*. Although Green is given authorship, it is published with A. S. Collins, his partner at the Caledonia hatchery. Green would sell the hatchery to Collins, and in 1875 it was acquired by the State of New York.
38. Roosevelt and Green, *Fish Hatching and Fish Catching*, 10.

39. For an overview of Stone's early life, I have relied on Hedgpeth's "Founders of Fish Culture" and "Livingston Stone and Fish Culture," 1–22.
40. Stone recounts his early experience and provides a brief history on the topic in his "Artificial Propagation of Salmon," 205–35.
41. Stone was not alone in his attempts to spawn Atlantic salmon. Samuel Wilmot first tried to artificially propagate salmon in Canada in the 1860s, though it was not until 1870 that he mastered the process. The first salmon hatchery in Canada was built in 1870 near the village of Newcastle, Ontario. In the United States, Charles G. Atkins tried to propagate, though unsuccessfully, Atlantic salmon in the U.S. in 1871 using live fish bought from local fishermen. See Stickney, *Aquaculture in the United States*, 13–14.
42. Both quotes are from Stone, *Domesticated Trout*, 262 and 264, respectively.
43. See, for example, Bowen, "History of Fish Culture," 71–94.
44. *Report of the Fish Commissioners of the State of Vermont, 1867*, 22–23.
45. *Report of the Fish Commissioners of the State of Vermont, 1867*, 24.
46. The original sources for these quotes are, for Michigan, Clark et al., "History and Evaluation of Regulations"; and Barkwell, *Report of the Board of Fish Commissioners*, for the Wyoming quote. They appear in Rahel, "Unauthorized Fish Introductions," 431.
47. Whelan et al., "Tracking Fisheries Through Time," 395.
48. The notion that the association was formed to establish methods for prominent hatchery owners to fix the price of trout and salmon eggs seems to stem from the account of the organization's founding by Fred Mather in his book *Fish Culture*. Mather suggests that creating the association did not help hatchery owners sell their eggs, so they instead sought government action to help protect their industry from competition. This and other claims by Mather were disputed in later publications. For more, see Hedgpeth, "Livingston Stone and Fish Culture," note 4. See also Bowen, "History of Fish Culture," 78.
49. *Proceedings of the American Fish Culture Association*, 3.
50. *Proceedings of the American Fish Culture Association*, 38–39.
51. *Proceedings of the American Fish Culture Association*, 11.
52. Page, originally from Maine, helped organize the Oquossoc Angling Association of Rangeley, Maine, in 1867. Along with other members of this club he helped enact state fishing regulations that limited the harvest of trout in the streams around Rangeley. He also would serve as the president of the American Fish Culturists' Association in 1882–83.
53. Briefly, such innovations often increased the chances of successfully impregnating the eggs and seeing them reach maturity. In addition, causes of disease were

discovered and remedies found. And, of course, continued attempts to hybridize fish for certain desirable characteristics would produce fish that not only grew faster but also became adaptable to a larger variety of conditions. For a readable and far-ranging treatment, see Nash, *History of Aquaculture*.

3. The U.S. Fish Commission

1. I rely on two excellent introductions to Baird's life and his role in establishing the Commission. These are Rivinus and Youssef, *Spencer Baird of the Smithsonian*; and Allard, *Spencer Fullerton Baird*.
2. Rivinus and Youssef, *Spencer Baird of the Smithsonian* is especially noteworthy.
3. These included the U.S.-Mexico Boundary Survey (1848–55); the Pacific Railroad surveys (1853–55); the North Pacific Exploring Expedition (1853–56); and the Northwest Boundary Survey (1857–61).
4. Dawes also was connected with the Credit Mobilier scandal: in 1868 he received several thousand shares of the company's stock as part of the Union Pacific's influence buying attempts.
5. The Benjamin quote and the vote tally are from Allard, *Spencer Fullerton Baird*, 82.
6. Forty-First Congress, sess. 3, res. 18–22, vol. 16, 1871, 593–94.
7. Taylor, makes this point in *Making Salmon*, chapter 3.
8. The resolution is accessible at the NOAA History website, at history.noaa.gov/legacy/act3.html.
9. Shaw, "Memoir of Spencer Fullerton Baird," 146.
10. Baird, *Report on the Condition of the Sea Fisheries*, XXXIII.
11. Baird, *Report on the Condition of the Sea Fisheries*, XXXIII.
12. Baird, *Report on the Condition of the Sea Fisheries*, XXXIV.
13. Baird, *Report on the Condition of the Sea Fisheries*, XXXIV–XXV.
14. Allard, *Spencer Fullerton Baird*, 100.
15. When Roosevelt and Green approached Baird in 1871 with a similar idea, he rejected it. Allard makes this point in *Spencer Fullerton Baird*, 124n36.
16. *Congressional Globe*, 1, 397.
17. Roosevelt, *Fish Culture*.
18. Roosevelt, *Fish Culture*, 14.
19. Roosevelt, *Fish Culture*, 13.
20. Roosevelt, *Fish Culture*, 13.
21. Roosevelt, *Fish Culture*, 13.
22. Roosevelt, *Fish Culture*, 13.
23. Roosevelt, *Fish Culture*, 13.

24. William Clift, "Important Events in Fish Culture," 6. The reference to "barren rivers" means any river with populations of "rough" fish. Roosevelt makes a similar pitch in his speech when he states, "There is no reason why the waters of the West should be less prolific than those of the East, provided the right species were introduced; and were trout, salmon, bass, shad, and sturgeon to take the place of catfish, pickerel, and suckers, the gain would be manifest." Of course, the attitude of the times showed little concern for the potentially damaging effects that such a mass introduction of non-native species would have on indigenous ones. This topic has been and continues to be debated as the U.S. Fish and Wildlife Service, and their state counterparts continue with programs that create and introduce new species of fish to new waters. For an example of this argument, see Richard Conniff, "Disaster at Yellowstone," *New York Times*, 14 May 2019, 23.
25. Allard, *Spencer Fullerton Baird*, 129.
26. Baird to Senator Edmunds, in Allard, *Spencer Fullerton Baird*, 129.
27. Ludlow, *Report of a Reconnaissance*, 16. The quote is taken from Harris, "'No Finer Trout Streams in the World,'" 275–304. As Harris notes, "Within forty years, the watercourses of the Black Hills would be reengineered into an organic machine in which two million trout annually were hatched, distributed, and protected from unsustainable harvesting through the cooperative efforts of individual landowners, railroad companies, state game wardens, federal foresters, and hatchery workers" (275).
28. Reiger, *American Sportsmen*, 201.

4. The Fish Commission

1. Edmunds, "Introduction of Salmon," 38–39.
2. I do this to help manage the extent of the discussion. One reason not to include it is because it was limited in geographic scope. The Commission's project to propagate Atlantic salmon (and a few other salmon species) focused primarily on the East Coast.
3. One reason for this interest in eastern fish comes from the fact that many of the California commissioners were from the East Coast and were curious whether species they were familiar with could be brought to California. Another species of interest to them was the Eastern brook trout.
4. Commissioner of Fisheries of the State of California, *Biennial Report, 1874–75*, 19; quoted in Towle, "Authored Ecosystems," 63.
5. Letter to Livingston Stone, 31 May 1873; cited in Taylor, *Making Salmon*, 76. Not only did Baird attempt to have the Commission's program reach every

Congressional district, but he also extended the reach of the Commission by shipping fish eggs, notably salmon and rainbow trout, to far-flung places like Australia and New Zealand. Such efforts would help raise the stature of the U.S. scientific community in the world's eyes.

6. The source for this discussion and the quotes is Clift's diary of his trip found in the section of the commissioner's "Report on Shad-Hatching Operations: Operations in 1872." This is in *Report of the Commissioner for 1872 and 1873*, 403–5.
7. Baird, *Report of the Commissioner for 1873–74 and 1874–75*, xvii–xviii.
8. Both quotes are from Baird, *Report of the Commissioner for 1873–74 and 1874–75*, xvi.
9. Quotes are from Stone, "Artificial Propagation of Salmon," 206.
10. Livingston to Spencer Baird, 18 June 1872, in Towle, "Authored Ecosystems," 61–62. Towle provides an overview of Stone's work in California, extending beyond his oversight of the salmon operation.
11. For some perspective, if Stone had traveled to Omaha by rail and taken the stagecoach to California, it would have taken twenty-five days to reach San Francisco, assuming no unforeseen glitches. If he had taken an ocean voyage from Boston to San Francisco, it would have been even more time consuming and dangerous. As one advertisement of the day hyped, the transcontinental railroad was faster and allowed travelers to avoid "The Dangers of the Sea!" It took Stone a few days to reach Omaha, with the rest of the journey to San Francisco lasting about a week.
12. Stone built a working relationship with the Wintu, to the point where he even looked after their sick. Though he praised them for their hard work, he also complained about their constant pilfering of small items.
13. For more on the salmon decline in this era, see Towle, "Great Failure."
14. This was a variation on the standard egg-packing scheme. In his book *Domesticated Trout: How to Breed and Grow Them* (1873) Stone describes a very similar approach used by Seth Green wherein trout eggs are packed in round tins between alternating layers of moss. In fact, the French used a very similar process much earlier.
15. Stone, "Report of Operations," 421–23.
16. I am assuming that the trip from Redding to Sacramento was on the Central Pacific Railroad Company: Stone's account is not clear on this. Also, where the Union Pacific express car was shunted from the Michigan Central to the Great Western Railroad is not made clear.
17. This and the previous quote are from Stone, "Report of Operations," 422.
18. Stone, "Report of Operations," 449.

19. In the mid-1960s the Michigan fish commission successfully created a salmon fishery by introducing Coho salmon into the Platte River from a state hatchery located near the town of Honor, Michigan. From there the salmon migrated down to Lake Michigan and returned to spawn. For a firsthand accounting of the early days of the salmon project in Michigan, see Tanner and Tody, "History of the Great Lakes Salmon Fishery," 139–53.
20. Later, hatcheries would be set up along several rivers in Oregon to catch salmon for the same purpose, except that these salmon fry would be released back into rivers of the Northwest. On this see Taylor, *Making Salmon*, chapter 6.
21. Stone, "Artificial Propagation of Salmon," 219.
22. Stone's foresight was such that he argued for leaving the entire McCloud River drainage in control of the local Native Americans—the Wintu—to help guarantee the abundance of the wild salmon. When his recommendation was ignored, he sought to create a "National Salmon Park" in Alaska. It is somewhat ironic that Stone complained that "not only is every contrivance employed that human ingenuity can devise to destroy the salmon of our West coast rivers, but more surely destructive, more fatal than all is the slow but inexorable march of those destroying agencies of human progress, before which salmon must surely disappear as did the buffalo of the plains and the Indian of California." See Stone, "National Salmon Park," 149–62.
23. Taylor, *Making Salmon*, provides an excellent history of the salmon fishery of the Northwest and vain attempts by the Commission to solve its population decline. I say vain attempts based on the observation that annual returns of salmon today range from 750,000 to 3 million compared to historical levels of annual returns from 10–16 million fish. Moreover, 80 percent of the recent migration originated from hatcheries. These figures are from Hill and Kolmes, "Resilience History of the Columbia River Basin," 10, 76.
24. Quoted in Stone, "Artificial Propagation of Salmon," 219.
25. I will leave it to those more qualified to debate exactly which kind of trout Stone shipped. Needham and Behnke, "Origin of Hatchery Rainbow Trout," 156–58, argue that Stone may have shipped eggs from trout that were not resident (monanadromous) or fine-scaled McCloud rainbows (what Stone called red-sided trout) but sea-run steelhead. Behnke, "Livingston Stone," 20–22, argues that most of the rainbows used as brood stock in hatcheries to which Stone sent eggs could have been some mixture of the two.
26. Some fifty-nine thousand eggs either failed to become fertilized or simply were unusable.
27. Ferguson, *Report of T. B. Ferguson*, 63.

28. Many members of the Ornithological and Piscatorial Acclimatization Society of California, formed in 1870, were transplants from the East. They were, therefore, acquainted with Eastern fish culturists like Seth Green. In the spring of 1875, the society shipped five hundred rainbow eggs from fish taken from the San Francisco Bay Area to Seth Green at his Caledonia hatchery. Green referred to these as "California mountain trout." Because the State of New York had purchased Green's hatchery by then, technically this 1875 shipment of rainbow eggs from California to New York makes it the first state to receive rainbow trout eggs from California.

29. Holberton, "Successful Stocking of Streams"; and Redmond, "On Rainbow Trout," 192 and 230, respectively.

30. Cited in Smiley, "Notes upon Fish and Fisheries," 104.

31. McDonald played an important role in the evolution of the Commission's approach to stocking rainbow. His belief that hatcheries like the ones at Northville (MI) and Wytheville (VA) should be the Commission's sole suppliers of trout was based on his "engineering" approach to fish culture: fish could be produced like any other product. His fondness for Wytheville stems from the fact that as a member of the Virginia Fish Commission he lobbied for Wytheville as one of the few federal hatcheries in the country. For more on McDonald's background and life, see Marshall McDonald, https://www.seafareproject.eu/marshall-mcdonald/ (accessed 17 April 2019).

32. McDonald, "Report on Distribution of Fish and Eggs," 388.

33. McDonald had a significant role in setting Commission policy by this time. Baird was in ill health and relegated much of the operational decision-making to McDonald. Following Baird's death in August 1887, President Grover Cleveland appointed McDonald to the position of commissioner in January 1888. The relationship between McDonald, Stone, and apparently much of the Commission staff was strained. Once McDonald became commissioner he showed little appreciation for Stone's services, trying several times to remove him from the Commission's payroll. The sad end to Stone's lustrous career with the Commission is dealt with in Taylor, *Making Salmon*, 83–85; and Halverson, *Entirely Synthetic Fish*, 45–46.

34. Why is it so popular? Rainbow trout have the quality of being hybridized to create a fish with desired qualities. If you wanted a trout that spawned early or late, grew faster than "normal," or withstood larger temperature variations, it could be supplied. The rainbow's characteristics are so malleable that the National Fish Strain Registry lists more than seventy-five strains of rainbow trout. See Halverson, *Entirely Synthetic Fish*, 78.

35. One of the most successful of these entrepreneurs was Julius A. Poppe. The story is that he took delivery of five (yes, only five) carp at his California farm in 1872 and within a few years had enough propagated to open a successful business. Because of the ease with which Poppe propagated carp, he set about to advertise his success. Any farmer with a pond, Poppe advised, could raise carp to sell like any other crop. See National Park Service, "History of Common Carp in North America," nps.gov (accessed 15 June, 2025).
36. Doughty, "Wildlife Conservation," 174.
37. Baird, *Report of the Commissioner for 1877*, 42.
38. The arrival of carp in America was an outcome that aligned personal interests of two fish culturists. Sandiford suggests that "one [Baird] was a politically astute biologist who had just been appointed chief of the U.S. Fish Commission (USFC), the other [Hessel] was a fish culturist from Germany looking to build a new life in America. They both had agendas, and they each held a key to the other person's agenda." Sandiford, *Transforming an Exotic Species*, ix.
39. McDonald, "Distribution of German Carp," 94.
40. For a discussion, see Varley and Schullery, *Yellowstone Fishes*.
41. Sandiford, *Transforming an Exotic Species*, ii.
42. The public's fascination with carp was such that on 8 February 1884 the American Carp-cultural Association was formed in Philadelphia. As reported in the *New York Times*, "Its principal object is the dissemination of useful and trustworthy information among its members." "To Aid Carp Culture," *New York Times*, 9 February 1884, 5.
43. *Report of the Missouri Fish Commission, 1885–86*, 57.
44. Sandiford, *Transforming an Exotic Species*, ii.
45. The origins of these fish tell an interesting story: The so-called German brown trout that was being planted in streams around the country might actually have originated in Scotland. The same confusion about the carp arose depending on when the fish were imported. It is also interesting that the American fish culturists focused on transplanting native species and did not, at this time at least, venture far into the introduction of foreign species. They did, however, have no qualms about shipping rainbow trout around the globe. On these issues see Towle, "Authored Ecosystems"; and Halverson, *Entirely Synthetic Fish*.
46. For example, the range of the Eurasian tree sparrow in the United States is limited to only the St. Louis area.
47. Over two dozen species of North American fishes have become extinct, largely due to the introduction of nonnative species. The causes stem from competition

over limited food resources, predation, and the spread of new diseases among the native fish population. See Miller et al., "Extinctions of North American Fishes," 22–38.

48. Baird, *Report of the Commissioner for 1872 and 1873*, 749.
49. Baird, *Report of the Commissioner for 1872 and 1873*, 752.
50. Baird, *Report of the Commissioner for 1872 and 1873*, 752.
51. Baird, *Report of the Commissioner for 1872 and 1873*, 751.
52. Baird, *Report of the Commissioner for 1872 and 1873*, 751.
53. Baird, *Report of the Commissioner for 1877*, 25.
54. George H. Jerome, "Preparations for Stocking Michigan Lakes with Whitefish," *Hillsdale Standard*, 2 February, 1876, 1. Jerome warned that depositors "must come to time, meet the cans on their arrival, though the heavens fall. Should sickness or other calamity befall the regular depositor, a good substitute must be supplied, else the fish will be dumped to the platform."
55. Cited in Ferguson, *Report of T. B. Ferguson*, 72.
56. Ferguson and Downs, *Report of the Commissioners*, 52.
57. Commissioners of Fisheries, Game, and Forests, *First Annual Report*, 22.
58. *Report of the Fish Commission of the State of Missouri for 1885–86*, 17. They also could be punishing: in another report, the commission stated that assistance came from "all the railroad companies with only one exception, (that of the Chicago & Alton R.R.) when called upon." Makes you wonder if the Chicago & Alton Railroad fell into line after this public shaming.

5. The Fish Car

1. Stone's recounting of the accident is from "Operations in California in 1873," 385–90.
2. Quoted in Raymond, "Livingston Stone," 18–22.
3. Baird, *Report of the Commissioner for 1881*, XIV.
4. Baird, *Report of the Commissioner for 1885*, XXXVI.
5. The plan is detailed in McDonald, "Report of Distribution of Carp," 1121–26.
6. McDonald, "Report of Distribution of Carp," 1125.
7. "More Carp: The Arrival of the United States Fish Car," *Dallas Herald*, 12 January, 1882, 5.
8. "The Success of Carp Culture," *New York Times*, 30 October, 1884, 8.
9. *Opelousas Courier*, 7 January, 1882, 1.
10. See Doughty, "Wildlife Conservation," 169–96.
11. McDonald, "Report of the Division of Distribution," 1056.

12. Eastman, "Description of the United States Fish Commission Car No. 2," 39–41. Eastman did not live to see the success of his fish car: He died the same year that his fish car was introduced.
13. Baird, *Report of the Commissioner for 1882*, xci–xcii.
14. My source for this article is *Forest and Stream*, 13 September 1883, 128. The fact that the *Tribune* bothered to report on the Commission's newest fish car is just another example of the public's interest in the Commission's activities.
15. Leonard, *Federal Fish Car Era*.
16. *Report of Board of Illinois State Fish Commissioners, 1888*, 5.
17. Fish Commission of Nebraska, *Tenth Annual Report*, 15.
18. Fish Commission of Nebraska, *Tenth Annual Report*, 15.
19. Fish Commission of Nebraska, *Tenth Annual Report*, 15.
20. Fish Commission of Iowa, *Twelfth Biennial Report*, 8.
21. *Forest and Stream*, 18 August 1881, 53.
22. Because it is outside the scope of our time frame, Kansas took delivery of its fish car in 1906 and Montana in 1910.
23. *Report of Board of Illinois State Fish Commissioners, 1888*, 29.
24. "Fish for Everybody," [Waterloo IA] *Courier*, 16 November 1897.
25. *Milford* [PA] *Dispatch*, 5 May 1892.
26. Springer, *America's Bountiful Waters*, 6.

6. Railroads, the Landscape, and Sport Fishing

1. Aron, *Working at Play*, 29.
2. The *Encyclopedia Britannica* defines tourism as "the act and process of spending time away from home in pursuit of recreation, relaxation, and pleasure."
3. Aron, *Working at Play*, 50.
4. Rothman, *Devil's Bargains*, 43. Shaffer, *See America First* is a key source in this area of study.
5. "Some Western Resorts," *Harper's New Monthly Magazine*, August 1882, 325.
6. Athearn, *Westward the Briton*, 116; quoted in Morin, *Frontiers of Femininity*, 26.
7. The revenue data are from U.S. Bureau of the Census, *Historical Statistics of the United States, Colonial Times to 1957*, 428. The other data are from Middleton et al., *Encyclopedia of North American Railroads*, appendix A, 1134, 1138.
8. Stilgoe, *Train Time*, 108.
9. Austin, *Catskill Rivers*, 25. The first edition was published in 1983.
10. For more on advertising in this era, see Presbrey, *History and Development of Advertising*, especially chapter 47, "Early Days of Transportation Advertising."

11. For an example of this argument, see Schivelbusch, *Railway Journey*. This view hearkens back to the argument put forth by Leo Marx in *Machine in the Garden*.
12. For a broader discussion of railroads' influence on American landscape art, see Stilgoe, *Metropolitan Corridor*.
13. "Artists' Railroad Excursion," 4.
14. "Artists' Railroad Excursion," 12.
15. Babcock, *Our American Resorts*, vii.
16. This copy along with a beautiful view of the Zephyr snaking its way through the mountains can be found in an online Amtrak ad. See https://www.amtrak.com/routes/california-zephyr-train.html (accessed 29 July 2025).
17. Richter, *Home of the Rails*, 20.
18. Schullery, *American Fly Fishing*, 131.
19. Runte, *Trains of Discovery*, 5.
20. Arthur Reed Kimble, "A Chance for Scenery," *Outlook*, 26 November, 1898, 773. Interesting that by this time the once revered railroad industry was now lumped into the category of "soulless" corporations.
21. Stilgoe, *Metropolitan Corridor*, 229, 230.
22. Orsi, *Sunset Limited*, 196.
23. Runte, *Allies of the Earth*, 4.
24. A general sense of how these resorts fit into the history of tourism can be found in Shaffer, *See America First*; or Aron, *Working at Play*. For each resort mentioned, one can find a specific book. I should also mention that there were several resorts outside of the Northeast that attracted visitors during the antebellum period. These include White Sulfur Springs, several springs in Virginia, and Catoosa Springs in Georgia, among others.
25. Squeri, *Better in the Poconos*, 11.
26. See Dye, *All Aboard for Santa Fe*.
27. Quoted in Stilgoe, *Train Time*, 109.
28. Sopko, "Amusement Parks," 1.
29. A more complete discussion of Schlict's Mill and the role of the Frisco in popularizing this location as a fishing and tourist destination can be found in Morrow, "Before Bass Pro," 1–23.
30. Schullery, *If Fish Could Scream*, 103–8. Chapter 6 of Schullery's book provides a good introduction to the history and the debate over the creation of Yellowstone as a national park and the role of railroads. See also Runte, "National Parks," in *Encyclopedia of North American Railroads*, 734–37.
31. "The Yellowstone Park," *New York Times*, 12 March 1873, 4. It is interesting to note that in the early debates one argument for setting aside the large tracks

of land that would become national parks was that the land was worthless for commercial development.
32. This quote appears in Runte, "National Parks," 734. "Soulless" seems to have been a popular adjective for railroads in those days.
33. Ross-Bryant, *Pilgrimage*, 7. Even today if you respond negatively to the question "Have you been to [fill in a major national park, say the Grand Canyon]?," the response is often one of "Really?," as if you have not performed some act of expected public service.
34. Shaffer, *See America First*, 95.
35. For more on this, see Runte, *Allies of the Earth*.
36. Burns, *National Parks*; quoted in Ross-Bryant, *Pilgrimage*, 17.
37. This role of railroads and how railroads preserved these and other areas of wilderness is discussed in Runte, *Trains of Discovery*, 15.
38. Runte, *Trains of Discovery*, 5.
39. Shaffer, *See America First*, 20.
40. Anglers had long been aware of fishing in the East and later in the West. Railroads now allowed them to actually visit these locations. To illustrate, consider the titles and dates of this representative sample of articles taken from Missouri newspapers: "Fall Life in the Adirondacks: Deer Shooting and Trouting in the Great Forest," *Daily Missouri Republican*, 25 October 1864; "Hints for Male Tourists: Chiefly in Relation to Catching Trout," *St. Louis Republican*, 11 July 1875; and "A Day's Fishing in Devil's Gulch [Colorado]," *St. Joseph Daily Herald*, 6 September 1891.
41. Then former president Ulysses S. Grant hammered home a golden spike near Gold Creek Montana to celebrate the event.
42. Wheeler, *6000 Miles*.
43. Quoted in Brown, *Trout Culture*, 70.
44. The company quotes are found in Schwantes, "Tourists in Wonderland," 7.
45. For example, a poster for the Monon Route shows the route from northern Indiana to the Gulf Coast overlaid on an alligator. The Northwestern Railroad has a pointer overlaying its map of the western half of the country with its tail pointing to St. Paul (the home office) and its nose nudging up against Seattle. One railroad's map, for whatever reason, is overlaid onto a long-handled fry pan. In each instance, the idea was first to attract attention and then illustrate where the train went.
46. Unfortunately, the combination of easier access, lack of enforced regulations, and the mentality of the general angler of the times led to Michigan's native grayling being fished out early into the 1900s. Perhaps it is ironic, then, that Trout

Unlimited, a conservation group whose goal is the preservation of native fish and their environment, was formed in Michigan during the mid-1900s. A history of the Grand Rapid and Indiana Railroad is found in Meints, *Fishing Line*.

47. This discussion draws on Hafer, "Railroad Advertising."
48. I have expanded on this theme in Hafer, "Railroad Advertising."
49. Mershon, "Michigan Trout," 8–11. This article is an excerpt from his 1923 book, *Recollections of My Fifty Years Hunting and Fishing*.
50. Detroit, Lansing & Northern Railroad Company, *Detroit and the Pleasure Resorts of Northern Michigan*, n.p.
51. Quoted in Miller, *Old Au Sable*, 29.
52. This discussion draws on Hafer, *From Northern California*.
53. *Forest and Stream*, 11 February 1892, 127.
54. More detail about the Salem Line's stocking activities and the Current River can be found in Hafer, *From Northern California*.
55. Brown, *Trout Culture*, 68.
56. This incentive program is described in Schullery, *American Fly Fishing*, 131.
57. These figures are taken from George Alfred Townsend, "Pullman," *Weekly Chillicothe* [MO] *Crisis*, 22 June 1882, 1.
58. These latter two examples are discussed, and source materials cited, in Brown, *Trout Culture*, 71.
59. See Borgelt, "Flies Only," 112–13. For more on inspection cars, see Goldfeder, "Inspection Locomotive."

7. Railroads, Women, and Fishing

1. What to call a woman who fishes? "Fisherwoman" or the gender-neutral alternatives "fisher" or "angler"? For an interesting examination of this lexicographic issue, see Ogden, "Fisherwomen," 111–17.
2. For a brief history of the Pullman Company, among others, see Middleton, "Pullman Co."
3. These quotes are from Pomeroy, *In Search of the Golden West*, 123.
4. This quote is from Quinzio, *Food on the Rails*, 34.
5. For a history of the Harvey Girls and their relationship with the railroads, see Poling-Kempes, *Harvey Girls*.
6. The quote is from Enss, *Iron Women*, 137.
7. Richter, *Home of the Rails*, 60.
8. I am aware of the classist and discriminatory aspects of these ads: many women could not afford to travel in the manner of those shown, and those that were shown uniformly were young and white. Before dismissing the railroads as perpetuating

classism, ageism, and racism, don't such characterizations also apply to many of the advertisements seen on television today?

9. Chicago & North-Western Railway Company, *The North and West Illustrated for Tourists, Business and Pleasure Travel*, 117; accessed at HathiTrust, https://hdl.handle.net/2027/loc.ark:/13960/t53f52136.
10. Philadelphia & Reading Railway Co., *Rules of the Operating Department* (1903), 119.
11. Quoted in Richter, *Home on the Rails*, 93.
12. An ugly side of the business was the creation of the Jim Crow car, which had a partition that separated white and Black passengers. The nonwhite section of the car did not have a toilet or a water fountain. It also was common for Black women passengers to be excluded from the ladies' car and were often told to sit in the smoking car. Such behavior also led to two lawsuits. In 1871 Mary Jane Chilton, who was denied access to the ladies' car, sued the St. Louis, Iron Mountain & Southern Railroad for five thousand dollars. The local court found in favor of the railroad. When appealed to the Missouri Supreme Court, they, too, found in favor of the railroad. In 1881 a Black family sued the Cincinnati Southern Railroad for a similar reason. In this case the family had purchased first-class tickets on the line but was denied access to their seats on the basis of their color. See Enss, *Iron Women*, 94–96.
13. I am cheating a bit here because the Phoebe Snow campaign began in the early 1900s and ended in 1917. Though the Snow campaign is just beyond my 1900 cutoff, it was so influential and fits the story of how railroads marketed to women that it must be included in this discussion. Indeed, some of the advertising discussed below also is outside my arbitrary cutoff but are included for the same reason.
14. *Ladies Home Journal*, January 1890, 16.
15. Brown, *Trout Culture*, 17.
16. *Harper's Bazaar* featured many articles that promoted changes associated with the New Woman movement. Its column "The Outdoor Woman" ran for many years, dealing with various activities that women could engage in. For a useful cataloging of the many articles in this vein, see White, *Representations of the True Woman*.
17. McMurray, *Rod of Her Own*, 99–100.
18. These are reprinted in "Ladies in Camp."
19. F. Cauthorn, "Feminine Success on Trouting," *Forest and Stream*, 6 April 1876, 143.
20. "Fisherwoman," *Forest and Stream*, 18 January 1894, 81.
21. Townsend, "Woman's Trout-Fishing."

22. Paul Schullery, quoted in Brett French, "First Yellowstone Fishing Tale," *Billings Gazette*, 24 March 2021, https://afftafisheriesfund.org/blog-history/2014/1/23/get-out-there-smzce.
23. McMurray, "Rivaling the Gentleman," 106.
24. For more on Crosby's life, see, among others, Hogan, "Glamour Girl"; and Verde, "Dianna of the Rangeleys." Others tell this story, but the basics are the same.
25. Cruikshank, "Woman's Standpoint," 124.
26. Additional examples can be found in Hafer, "Trains and Trout," 10–23.
27. The *Literary Digest* was a popular general readership weekly with a national circulation of about 1,400,000. Each issue covered topics ranging from foreign policy and national politics to religion, science, and the arts. The digest proclaimed itself to be the magazine for the "alert" woman.
28. *Munsey's Magazine* began as *Munsey's Weekly* in 1889. In 1891 it switched to monthly publication. The magazine, like the *Literary Digest*, was a general-interest publication, its circulation hovering around seven hundred thousand by the late 1890s. Advertising in *Munsey's*, the *Literary Digest* and other such publications indicates the railroads' attempt to reach a broad audience that included urban and rural readers of both sexes.
29. The series of posters that Welsh created shows women engaged in a variety of sporting activities, such as skiing and sailing. The women all appear to be quite fashionable, which had the dual effect of a) appealing to the wealthier female traveler and b) suggesting to others that if they rode in a Pullman car and visiting these vacation sites, they, too, might be considered fashionable. As many railroad ads suggest, you're known by the company you keep.
30. See the argument as presented in Waring, *Counting for Nothing*. Indeed, in economics a long-standing debate involves how to accurately measure the market value of women's work around the house in order to get an accurate picture of a country's true output.
31. For a discussion of how women's position in the hierarchy of fly fishers has (or has not) changed over time, see Crowder, "Check Your Fly," 162–65.
32. Melner and Kessler, *Great Fishing Tackle Catalogs*, 46.
33. Melner and Kessler, *Great Fishing Tackle Catalogs*, 208.
34. In more modern times, such recognition was the catalyst for the creation of many conservation groups, including Trout Unlimited and Ducks Unlimited.
35. Reiger, *American Sportsmen*, 65.
36. Riley, *Women and Nature*, 145.
37. Fox makes this point in his *John Muir and His Legacy*.

38. Riley, *Women and Nature*, xiii. See also her "Victorian Ladies Outdoors" and the articles cited therein.
39. Riley, *Women and Nature*, xiii.
40. Cited in Riley, *Women and Nature*, 147.

One Final Note
1. Springer, *America's Bountiful Waters*.

BIBLIOGRAPHY

Allard, Dean Conrad, Jr. *Spencer Fullerton Baird and the U.S. Fish Commission.* New York: Arno, 1968.

Ambrose, Stephen E. *Nothing Like It in the World: The Men Who Built the Transcontinental Railroad, 1863–1869.* New York: Simon & Schuster, 2000.

Ames, Charles Edgar. *Pioneering the Union Pacific: A Reappraisal of the Builders of the Railroad.* New York: Meredith Corporation, 1969.

Atack, J., and P. Passell. *A New Economic View of American History: From Colonial Times to 1940.* New York: W. W. Norton, 1994.

Athearn, Robert G. *Westward the Briton.* Lincoln: University of Nebraska Press, 1962.

Aron, Cindy S. *Working at Play: A History of Vacations in the United States.* New York: Oxford University Press, 1999.

"Artists' Railroad Excursion over the Baltimore and Ohio Rail Road." *New Harper's Monthly Magazine* 9, no. 109 (June 1859): 1–19.

Austin, Francis. *Catskill Rivers: Birthplace of American Fly Fishing.* New York: Skyhorse, 2014.

Babcock, Louis. *Our American Resorts. For Health, Pleasure, and Recreation.* New York: National News Bureau, 1881.

Baird, Spencer F. *Report on the Condition of the Sea Fisheries of the South Coast of New England in 1871 and 1872.* Washington DC: GPO, 1873.

———. *Report of the Commissioner for 1872 and 1873.* Washington DC: GPO, 1874.

———. *Report of the Commissioner for 1873–74 and 1874–75.* Washington DC: GPO, 1876.

———. *Report of the Commissioner for 1877.* Washington DC: GPO, 1879.

———. *Report of the Commissioner for 1881.* Washington DC: GPO, 1884.

———. *Report of the Commissioner for 1885.* Washington DC: GPO, 1887.

Baker, George Piece. *The Formation of the New England Railroad System.* Cambridge MA: Harvard University Press, 1949.

Barkwell, M. C. *Report of the Board of Fish Commissioners for the Two Years Ending Dec. 31, 1883*. Cheyenne: Wyoming Game and Fish Department, 1883.

Barky, Ian R. "The Invention of Railroad Time." *Railroad History* 148 (Spring 1983): 12–22.

Behnke, Robert. "Livingston Stone, J. B. Campbell, and the Origins of Hatchery Rainbow Trout." *American Fly Fisher* 16, no. 3 (Fall 1990): 20–22.

Benson, Norman G., ed. *A Century of Fisheries in North America*. Washington DC: American Fisheries Society, 1970.

Bertram, James Glass. *The Harvest of the Sea: Contributions to the Natural and Economic History of the British Food Fishes*. Edinburgh: R. Clark, 1865.

Black, Sylvia R. "Seth Green: Father of Fish Culture." *Rochester History* 6, no. 3 (July 1944): 1–24.

Blondheim, Menahem. *News over the Wires: The Telegraph and the Flow of Public Information in America, 1844–1897*. Cambridge MA: Harvard University Press, 1994.

Borgelt, Bryon. "Flies Only: Early Sport Fishing Conservation on Michigan's Au Sable River." PhD diss., University of Toledo, 2009. https://etd.ohiolink.edu/acprod/odb_etd/ws/send_file/send?accession=toledo1242090675&disposition=inline.

Bowen, J. T. "A History of Fish Culture as Related to the Development of Fishery Programs." In *A Century of Fisheries in North America*, edited by Norman G. Benson, 71–94. Washington DC: American Fisheries Society, 1970.

Brown, Margaret L. "Asa Whitney and His Pacific Railroad Publicity Campaign." *Mississippi Valley Historical Review* 20, no. 2 (September 1933): 209–24.

Brown, Jen Corrine. *Trout Culture: How Fly Fishing Forever Changed the Rocky Mountain West*. Seattle: University of Washington Press, 2015.

Bryant, Keith L., Jr. "Development of North American Railroads." In *Encyclopedia of North American Railroads*, edited by William D. Middleton, George M. Smerk, and Roberta L. Diehl, 1–17. Bloomington: Indiana University Press, 2007.

Burd, Camden. "Imagining a Pure Michigan Landscape: Advertisers, Tourists, and the Making of Michigan's Northern Vacationlands." *Michigan Historical Review* 42, no. 2 (Fall 2016): 31–51.

Burns, Ken, dir. *The National Parks: America's Best Idea*. Florentine Films and WETA Television, 2009.

Butler, John A. "Some Western Resorts." *Harper's New Monthly Magazine* 65, no. 387 (August 1882): 325–41.

Caldwell, Charles, "On the Moral and Other Indirect Influences of Rail-Roads." *New England Magazine* 2 (January–June 1832): 288–300.

Chicago & North-Western Railway Company. *The North and West Illustrated for Tourists, Business and Pleasure Travel*. Chicago: Chicago & North-Western Railway Company, ca. 1880.

Clark, R. D., Jr., G. R. Alexander, and H. Gowing. "A History and Evaluation of Regulations for Brook Trout and Brown Trout in Michigan Streams." *North American Journal of Fisheries Management* 1 (1981): 1–14.

Clift, William. "Important Events in Fish Culture During the Year 1872." *Proceedings of the American Fish Culturists' Association*. Albany NY: Argus, 1873.

———. "Report on Shad-Hatching Operations: Operations in 1872." In *Report of the Commissioner for 1872 and 1873*, 403–5. Washington DC: GPO, 1874.

Commissioner of Fisheries of the State of California. *Biennial Report, 1874–75*. Sacramento: California Fish Commission, 1876.

Commissioners of Fisheries, Game, and Forests of New York. *First Annual Report*. New York: Wynkoop Hallenbeck Crawford, Printers, 1896.

Congressional Globe: Containing the Debates and Proceedings of the Second Session Forty-Second Congress; With an Appendix, Embracing the Laws Passed at that Session. Washington DC: Office of the Congressional Globe, 1872.

Cotterill, Robert. S. "National Railroad Convention in St. Louis, 1849." *Missouri Historical Review* 12 (1918): 203–15.

Crawford, Jay Boyd. *The Credit Mobilier of America: Its Origin and History*. Boston: C. W. Calkins, 1880.

Crowder, Rachel. "Check Your Fly: Tales of a Woman Fly Fisher." *Canadian Woman Studies* 21, no. 3 (2002): 162–65.

Cruikshank, Mrs. James A.,"The Woman's Standpoint." In *In the Maine Woods*, written and arranged by Fred H. Clifford, 104. Bangor ME: Bangor & Aroostook R. R., 1904.

Dartnell, George. *A Proposed Plan for a Rail Road Clearing House*. Buffalo NY: Clapp, Matthews, 1858.

———. *Report on the Gauge for the St. Lawrence and Atlanta Railroad*. Portland ME: Thurston, 1847.

DeBow, J. D. B. "Additional Remarks by the Editor of the Projected Southern and Northern Routes Across the Continent to the Pacific." *Commercial Review of the South and West* 4 (June 1847): 485–95.

Detroit, Lansing & Northern Railroad Company. *Detroit and the Pleasure Resorts of Northern Michigan: Compliments of Passenger Department*. Detroit: J. F. Eby, 1883.

Doughty, Robin W. "Wildlife Conservation in Late Nineteenth-Century Texas: The Carp Experiment." *Southwestern Historical Quarterly* 84, no. 2 (October 1980): 169–96.

Duboff, Richard B. *Accumulation & Power: An Economic History of the United States*. Armonk NY: M. E. Sharpe, 1989.

Dye, Victoria E. *All Aboard for Santa Fe: Railway Promotion of the Southwest, 1890s to 1930s*. Albuquerque: University of New Mexico Press, 2005.

Eastman, Frank S. "Description of the United States Fish Commission Car No. 2, Designed for the Distribution of Young Fish." In *Report of the Commissioner for 1882*, 39–41. Washington DC: GPO, 1884.

Edmunds, M. C. "The Introduction of Salmon into American Waters." *Transactions of the American Fisheries Society* 1, no. 1 (1872): 38–39.

Enss, Chris. *Iron Women: The Ladies Who Helped Build the Railroad*. Guilford CT: Twodot, 2021.

Ferguson, T. B. *Report of T. B. Ferguson, Commissioner of Fisheries of Maryland*. Hagerstown MD: Bell, Printers, January 1881.

Ferguson, T. B., and Philip W. Downs. *Report of the Commissioners of Fisheries of Maryland*. Annapolis MD: John F. Wiley, State Printer, 1875.

Fish Commission of Iowa. *Twelfth Biennial Report of the Fish Commission of the State of Iowa, 1896–97*. Des Moines IA: F. R. Conaway, State Printer, 1897.

Fish Commission of Nebraska. *Tenth Annual Report of the Fish Commission of Nebraska, 1888*. Omaha: Henry Gibson, State Printer, 1889.

Forest and Stream, various issues. New York: Forest and Stream.

Forty-First Congress. Sess. 3, res. 18–22, vol. 16, 1871, 593–94.

Fox, Stephen. *John Muir and His Legacy: The American Conservation Movement*. Boston: Little, Brown, 1981.

Garlick, Theodatus. *A Treatise on the Propagation of Certain Kinds of Fish*. Cleveland OH: Tho. Brown, 1857.

Goldfeder, Ron. "The Inspection Locomotive." *Railroad History* 206 (Spring–Summer 2012): 8–47.

Goodrich, Carter. *Government Promotion of American Canals and Railroads, 1800–1890*. Westport CT: Greenwood, 1960.

Green, Seth. *Trout Culture*. Rochester NY: Green & Collins, 1870.

Hafer, Rik W. *From Northern California to the Ozarks of Missouri: How Rainbow Trout Came to the Show-Me State*. Amazon, 2021.

———. "Railroad Advertising, Nature, and Sport Fishing in America: Part II." *American Fly Fisher* 51, no. 2 (Spring 2025): 2–12.

———. "Trains and Trout: Railroads and the Evolution of Sport Fishing in America." *Railroad History* no. 226 (Spring–Summer 2002): 10–23.

Haines, Michael R., and Robert A. Margo. "Railroads and Economic Development." In *Quantitative Economic History: The Good of Counting*, edited by Joshua L. Rosenbaum, 78–99. New York: Routledge, 2008.

Hallock, Charles. *Vacation Rambles in Northern Michigan*. Grand Rapids MI: Passenger Department of Grand Rapids & Indiana Railroad Company, 1878.

Halverson, Anders. *An Entirely Synthetic Fish: How Rainbow Trout Beguiled America and Overran the World*. New Haven CT: Yale University Press, 2010.

Harris, John R. "'No Finer Trout Streams in the World': The Making of a Recreational Fishery in the Black Hills Forest Reserve." *South Dakota History* 45, no. 4 (Winter 2015): 275–304.

Hedgpeth, Joel W. "Founders of Fish Culture: Livingston Stone." *Progressive Fish-Culturist* 8, no. 55 (1941): 11–14.

———. "Livingston Stone and Fish Culture in California." *California Fish and Game* 27, no. 8 (July 1941): 1–22.

Hill, Gregory M., and Stephen A. Kolmes. "Resilience History of the Columbia River Basin and Salmonid Species: Regimes and Policies." *Environments*, 2023. https://doi.org/10.3390/environments10050076.

Hogan, Austin S. "Glamour Girl of the Maine Lakes: Fly Rod's Reel Was of Solid Gold." *American Fly Fisher* 4, no. 4 (Fall 1977): 5–7.

Holberton, Wakeman. "The Successful Stocking of Streams with Trout." In *Bulletin of the United States Fish Commission for 1882*, 192. Washington DC: GPO, 1883.

Jerome, George. *Third Report of the Superintendent of the Michigan State Fisheries for 1877–78*. Lansing MI: W. S. George, State Printers and Binders, 1879.

Kinsey, Darin. "'Seeding the Water as the Earth': The Epicenter and Peripheries of a Western Aquacultural Revolution." *Environmental History* 11, no. 3 (July 2006): 527–66.

"Ladies in Camp." *American Fly Fisher* 2, no. 1 (Winter 1975): 10–13.

Leonard, John R. *The Federal Fish Car Era of the National Fish Hatchery System*. Washington DC: Department of the Interior, U.S. Fish and Wildlife Service, 1979.

Lewis, Oscar. *The Big Four: The Story of Huntington, Stanford, Hopkins, and Crocker, and of the Building of the Central Pacific*. New York: Knopf, 1938.

Ludlow, William. *Report of a Reconnaissance of the Black Hills of Dakota, Made in the Summer of 1874*. Washington DC: GPO, 1875.

Marsh, George Perkins. *Report, Made Under the Authority of the Legislature of Vermont, on the Artificial Propagation of Fish*. Burlington VT: Free Press, 1857.

Martin, Albro. *Railroads Triumphant: The Growth, Rejection, and Rebirth of a Vital American Force*. New York: Oxford University Press, 1992.

Marx, Leo. *The Machine in the Garden: Technology and the Pastoral Ideal in America.* New York: Oxford University Press, 1964.

Mather, Fred. *Fish Culture in Fresh and Salt Water.* New York: Field & Stream, 1900.

———. "The Influence of Railroads." *Transactions of the American Fisheries Society* (January 1896): 17–27.

McDonald, Marshall. "Distribution of German Carp by the United States Fish Commission." In *Bulletin of the United States Fish Commission, for 1882,* 94. Washington DC: GPO, 1883.

———. "Report of Distribution of Carp, During the Season of 1881–82, by the United States Fish Commission." In *Report of the Commissioner for 1881,* 1121–26. Washington DC: GPO, 1884.

———. "Report of the Division of Distribution of the United States Fish Commission for the Year 1883." In *Report of the Commissioner for 1883,* 1051–87. Washington DC: GPO, 1885.

———. "Report on Distribution of Fish and Eggs by the U.S. Fish Commission for the Season of 1885–86." In *Bulletin of the United States Fish Commission for 1886,* 385–94. Washington DC: GPO, 1886.

McMurray, David. "Rivaling the Gentlemen in the Gentle Art: The Authority of the Victorian Woman Angler." *Sport History Review* (2008): 99–126.

———. "A Rod of Her Own: Women and Angling in Victorian North America." Master's thesis, University of Lethbridge, Alberta, Canada, 2007. https://opus.uleth.ca/server/api/core/bitstreams/c1afd30c-d360-46cd-bb1d-eb04cd7ff378/content.

Meints, Graydon M. *The Fishing Line: A History of the Grand Rapids & Indiana Railroad.* East Lansing: Michigan State University Press, 2018.

Melner, Samuel, and Hermann Kessler. *Great Fishing Tackle Catalogs of the Golden Ages.* New York: Crown, 1972.

Mershon, William. "Michigan Trout." *American Fly Fisher* 7, no. 4 (Fall 1980): 8–11.

Middleton, William D. "Pullman Co." In *Encyclopedia of North American Railroads,* edited by William D. Middleton, George M. Smerk and Roberta L. Diehl, 881–84. Bloomington: Indiana University Press, 2007.

Middleton, William D., George M. Smerk, and Roberta L. Diehl, eds. *Encyclopedia of North American Railroads.* Bloomington: Indiana University Press, 2007.

Miller, Harzen L. *The Old Au Sable.* Grand Rapids MI: William B. Erdermans, 1968.

Miller, R. R., J. D. Williams, and J. E. Williams. "Extinctions of North American Fishes During the Past Century." *Fisheries* 14, no. 6 (1989): 22–38.

Million, John Wilson. "State Aid to Railroads in Missouri." *Journal of Political Economy* 3, no. 1 (December 1894): 73–97.

Minor, D. K. "The Railroad Journal." *American Railroad Journal and Mechanic's Magazine* 12, no. 2 (December 1884): 1.

Morin, Karen M. *Frontiers of Femininity: A New Historical Geography of the Nineteenth-Century American West*. Syracuse NY: Syracuse University Press, 2008.

Morrow, Lynn. "Before Bass Pro: St. Louis Sporting Clubs on the Gasconade River." *Missouri Historical Review* 9, no.1 (October 2004): 1–23.

Nash, Colin E. *The History of Aquaculture*. Hoboken NJ: Wiley & sons, 2010.

Needham, P. R., and R. J. Behnke, "The Origin of Hatchery Rainbow Trout." *Progressive Fish Culturist* 24, no. 4 (1962): 156–58.

Ogden, Lesley Evans. "Fisherwomen: The Uncounted Dimension in Fisheries Management." *BioScience* 67, no. 2 (February 2017): 111–17.

Orsi, Richard J. *Sunset Limited: The Southern Pacific Railroad and the Development of the American West, 1850–1930*. Berkeley: University of California Press, 2005.

Philadelphia & Reading Railway Company. *Rules of the Operating Department* (1903).

Poling-Kempes, Lesley. *The Harvey Girls: Women Who Opened the West*. New York: Marlowe, 1991.

Pomeroy, Earl. *In Search of the Golden West: The Tourist in Western America*. 2nd ed. Lincoln: University of Nebraska Press, 2010.

Poor, Henry V. *Manual of the Railroads of the United States for 1872–73*. New York: H. V. & H. W. Poor, 1868.

———. "Sketch of the Rise, Progress, Cost, Earnings, Etc. of the Railroads of the United States." In *Manual of the Railroads of the United States for 1868–69*, 9–32. New York: H. V. & H. W. Poor, 1868.

Presbrey, Frank. *The History and Development of Advertising*. Garden City NY: Doubleday, Doran, 1929.

Proceedings of the American Fish Culturist Association. Albany NY: Argus, 1872.

Quinzio, Jeri. *Food on the Rails: The Golden Era of Railroad Dining*. Lanham MD: Rowman & Littlefield, 2014.

Rahel, Frank J. "Unauthorized Fish Introductions: Fisheries Management of the People, for the People, or by the People?" *American Fisheries Society Symposium* 44 (2004): 431–43.

Raymond, Frank E. "Livingston Stone: Pioneer Fisheries Scientist, His Career in California." *American Fly Fisher* 16, no. 1 (1990): 18–22.

Redmond, Roland. "On Rainbow Trout Reared from Eggs Brought from California." In *Bulletin of the United States Fish Commission for 1882*, 230. Washington DC: GPO, 1883.

Reiger, John F. *American Sportsmen and the Origins of Conservation*. 3rd ed. Corvallis: Oregon State University Press, 2001.

Report of Board of Illinois State Fish Commissioners, 1888. Springfield IL: Springfield Printing Company, 1889.
Report of the Commissioner for 1885. Washington DC: GPO, 1887.
Report of the Commissioner for 1889–1891. Washington DC: GPO, 1893.
Report of the Commissioner for 1895. Washington DC: GPO, 1896.
Report of the Commissioner for 1900. Washington DC: GPO, 1901.
Report of the Fish Commission of the State of Missouri for 1885–86. Jefferson City, MO: Tribune Printing Company, State Printers and Binders, 1887.
Report of the Fish Commissioners of the State of Vermont 1867. Montpelier VT: Walton's Steam Printing Establishment, 1867.
Richter, Amy G. *Home of the Rails: Women, the Railroad, and the Rise of Public Domesticity*. Chapel Hill: University of North Carolina Press, 2005.
Riley, Glenda. "Victorian Ladies Outdoors: Women in the Early Western Conservation Movement, 1870–1920." *Southern California Quarterly* 83, no. 1 (Spring 2001): 59–80.
———. *Women and Nature: Saving the "Wild" West*. Lincoln: University of Nebraska Press, 1999.
Rivinus, E. F., and E. M. Youssef. *Spencer Baird of the Smithsonian*. Washington DC: Smithsonian Institution, 1992.
Roosevelt, Robert Barnwell. *Fish Culture Compared in Importance with Agriculture: Speech of Hon. Robert B. Roosevelt, of New York, in the House of Representatives, May 13, 1873*. Washington DC: F & J. Rives & Geo A. Bailey, 1872.
———. *Superior Fishing: Or the Striped Bass, Trout, and Black Bass of the Northern States*. New York: Carleton, 1865.
Roosevelt, Robert Barnwell, and Seth Green. *Fish Hatching and Fish Catching*. Rochester NY: Union and Advertiser Co.'s Book & Job Print, 1879.
Ross-Bryant, Lynn. *Pilgrimage to the National Parks: Religion and Nature in the United States*. New York: Routledge, 2013.
Rothman, H. *Devil's Bargains: Tourism in the Twentieth-Century American West*. Lawrence: University Press of Kansas, 1998.
Runte, Alfred. *Allies of the Earth: Railroads and the Soul of Preservation*. Kirksville MO: Truman State University Press, 2006.
———. "National Parks." In *Encyclopedia of North American Railroads*, edited by William D. Middleton, George M. Smerk and Roberta L. Diehl, 734–37. Bloomington: Indiana University Press, 2007.
———. *National Parks: The American Experience*. Lincoln: University of Nebraska Press, 1979.
———. *Trains of Discovery: Railroads and the Legacy of Our National Parks*. 5th ed. Lanham MD: Roberts Rinehart, 2011.

Sabin, Edwin L. *Building the Pacific Railway*. Philadelphia: J. P. Lippincott, 1919.

Sandiford, Glenn. "Transforming an Exotic Species: Nineteenth Century Narratives about Introduction of Carp to America." PhD diss., University of Illinois at Urbana-Champaign, 2009. https://www.proquest.com/openview/955a636fa835721b9916e5927ed9140a/1?pq-origsite=gscholar&cbl=18750.

Schivelbusch, Wolfgang. *The Railway Journey: Trains and Travel in the 19th Century*. New York: Urizen, 1978.

Schullery, Paul. *American Fly Fishing: A History*. New York: Nick Lyons, 1987.

———. *If Fish Could Scream: An Angler's Search for the Future of Fly Fishing*. Mechanicsburg PA: Stackpole, 2008.

Schwantes, Carlos A. "Tourists in Wonderland: Early Railroad Tourism in the Pacific Northwest." *Columbia Magazine* 7, no. 4 (Winter 1993–94): 22–33.

Shaffer, Marguerite S. *See America First: Tourism and National Identity, 1880–1940*. Washington DC: Smithsonian Institution, 2011.

Shanks, W. F. G. "Fish-Culture in America." *Harper's New Monthly Magazine* 37, no. 222 (November 1868): 721–39.

Shaw, John Billings. *Memoir of Spencer Fullerton Baird, 1823–1887*. Washington: Judd & Detweiler, 1889.

Smiley, Chas. W. "Notes upon Fish and Fisheries." In *Bulletin of the United States Fish Commission for 1885*, 104. Washington DC: GPO, 1885.

Sopko, Jennifer. "Amusement Parks: A Confluence of Transportation, Industry and Topography." *Ephemera Journal* 25, no. 1 (September 2022): 1, 6–11.

Springer, Craig, ed. *America's Bountiful Waters: 150 Years of Fisheries Conservation and the U.S. Fish & Wildlife Service*. Guilford CT: Stackpole, 2021.

Squeri, Lawrence. *Better in the Poconos: The Story of Pennsylvania's Vacationland*. University Park: Penn State University Press, 2002.

Steedman, I. G. W. *Carp and Carp Culture in Missouri with Appendix on Native Fish*. 2nd ed. Jefferson City: Missouri Fish Commission, 1887.

Stickney, Robert B. *Aquaculture in the United States: A Historical Survey*. New York: John Wiley & Sons, 1996.

Stilgoe, John R. *Metropolitan Corridor: Railroads and the American Scene*. New Haven CT: Yale University Press, 1983.

———. *Train Time: Railroads and the Imminent Reshaping of the United States Landscape*. Charlottesville: University of Virginia Press, 2007.

"The St. Louis Railway Commenced." *American Railroad Journal* 25 (1852): 142.

Stone, Livingston. "The Artificial Propagation of Salmon on the Pacific Coast of the United States, with Notes on the Natural History of the Quinnat Salmon." In *Bulletin of the United States Fish Commission for 1896*, 203–25. Washington DC: GPO, 1897.

---. *Domesticated Trout: How to Breed and Grow Them*. Boston: James R. Osgood, 1873.

---. "A National Salmon Park." *Transactions of the American Fisheries Society* 21, no. 1 (1892): 149–62.

---. "Operations in California in 1873." In *Report of the Commissioner for 1873–74 and 1874–75*, 311–429. Washington DC: GPO, 1876.

---. "Report of Operations During 1874 at the United States Salmon-Hatching Establishment on the McCloud River, California." In *Report of the Commissioner for 1873–74 and 1874–75*, 437–78. Washington DC: GPO, 1876.

---. "Some Brief Reminiscences of the Early Days of Fish-Culture in the United States." In *Bulletin of the United States Fish Commission for 1897*, 337–43. Washington DC: GPO, 1898.

Stover, John F. *American Railroads*. Chicago: University of Chicago Press, 1961.

---. *Iron Road to the West: American Railroads in the 1850s*. New York: Columbia University Press, 1978.

Tanner, Howard A., and Wayne H. Tody. "History of the Great Lakes Salmon Fishery: A Michigan Perspective." In *Sustaining North American Salmon: Perspectives Across Regions and Disciplines*, edited by Kristine D. Lynch, Michael L. Jones, and William W. Taylor, 139–53. Bethesda MD: American Fisheries Society, 2002.

Taylor, George Roger, and Irene D. Neu. *The American Railroad Network: 1861–1890*. Cambridge MA: Harvard University Press, 1956.

Taylor, Joseph E., III. *Making Salmon: An Environmental History of the Northwest Fisheries Crisis*. Seattle: University of Washington Press, 1999.

Thies, Clifford F. "The American Railroad Network during the Early 19th Century: Private Versus Public Enterprise." *Cato Journal* 22, no. 2 (Fall 2002): 229–61.

Towle, Jerry C. "Authored Ecosystems: Livingston Stone and the Transformation of California Fisheries." *Environmental History* 5, no. 1 (January 2000): 57–74.

---. "The Great Failure: Nineteenth-Century Dispersals of the Pacific Salmon." *California Geographical Society* 27 (1987): 75–96.

Townsend, Mary Trowbridge. "A Woman's Trout-Fishing in Yellowstone Park." *Outing Magazine*, May 1897, 163–65.

---. "Angling for Eastern Trout." *Outing Magazine*, May 1899, 124–28.

Trelease, Allen W. "The Passive Voice: The State and the North Carolina Railroad, 1849–1871." *North Carolina Historical Review* 61, no. 2 (April 1984): 174–204.

Twelfth Biennial Report of the Fish Commission of the State of Iowa, 1896–97. Des Moines IA: F. R. Conaway, State Printer, 1897.

U.S. Bureau of the Census. *Historical Statistics of the United States, Colonial Times to 1957*. Washington DC: GPO, 1960.

———. *Tenth Census of the United States, Report on the Agencies of Transportation in the United States*. Washington DC: GPO, 1883.

U.S. Federal Highway Administration. *America's Highways: 1776–1796*. Washington DC: U.S. Department of Transportation, 1977.

Varley, John and Paul Schullery. *Yellowstone Fishes: Ecology, History, and Angling in the Park*. Mechanicsburg PA: Stackpole, 1998.

Verde, Thomas A. "Dianna of the Rangeleys." *American Fly Fisher* 15, no. 2 (1989): 9–13.

Vernon, Edward, ed. *Travelers' Official Railway Guide of the United States and Canada*. New York, June 1868.

Waldman, John. *Running Silver: Restoring Atlantic Rivers and Their Great Fish Migrations*. Guilford CT: Lyons, 2013.

Wallis, John Joseph. "What Caused the Crisis of 1839?" National Bureau of Economic Research, Historical Paper 133, April 2001.

Ward, James A. *Railroads and the Character of America: 1820–1887*. Knoxville: University of Tennessee Press, 1986.

Waring, M. *Counting for Nothing: What Men Value and What Women Are Worth*. Auckland: Auckland University Press, 1966.

Watt, William J. "Evolution of Major Railroads." In *Encyclopedia of North American Railroads*, edited by William D. Middleton, George M. Smerk and Roberta L. Diehl, 437–41. Bloomington: Indiana University Press, 2007.

Webster, Daniel. "Opening of the Northern Railroad." In *The Works of Daniel Webster*, Vol. 2. 6th ed., 414–19. Boston: Little Brown, 1853.

Weiss, Thomas. "Tourism in America Before World War II." *Journal of Economic History* 64 no. 2 (June 2004): 289–327.

Wheeler, Oiln D. *6000 Miles Through Wonderland: Being a Description of the Marvelous Region Traversed by the Northern Pacific Railroad*. St. Paul MN: Northern Pacific Railway, 1893.

Whelan, Gary E., Diana M. Day, and John M. Casselman. "Tracking Fisheries Through Time: The American Fisheries Society as a Historical Lens." *Fisheries* 45, no. 8 (August 2020): 392–426.

White, Erica M. "Representations of the True Woman and the New Woman in *Harper's Bazaar*, 1870–1879 and 1890–1905," Master's thesis, Iowa State University, 2009. https://www.proquest.com/openview/d479a03e24d14e103f65133a35dbb5cb/1?pq-origsite=gscholar&cbl=18750.

White, John H., Jr. *The American Railroad Passenger Car, Part 1 and 2*. Baltimore MD: Johns Hopkins University Press, 1978.

White, Richard. *Railroaded: The Transcontinentals and the Making of Modern America*. New York: W. W. Norton, 2011.

Wilson, F.W. *Journal of Capt. Daniel Bradley—An Epic of the Ohio Frontier with Copious Comments by Frazier E. Wilson*. Greenville OH: Frank H. Jones & Son, 1935.

Zega, Michael E. and John E. Gruber. *Travel by Train: The American Railroad Posters, 1870–1950*. Bloomington: University of Indiana Press, 2002.

INDEX

Abercrombie and Fitch, 166
Acadia National Park, 140
accidents (railroad), 103–7, 116, 121
Ackley, H. A., 33–34
Act of 1864, 10, 15
Adams Express Company, 74, 96
advertising (railroad industry): classist and discriminatory aspects of, 165–66, 194n8; encouraging tourism, xv–xvi, 127, 128–32, 135–36; featuring Phoebe Snow, 158–59, 195n13; "fishing line" campaigns, 143–45; guidebooks, 132, 141–43, 146; highlighting safety, 19, 20, 156, 157–58, 165, 169, 173; impacting sport fishing, xv–xvi, 20, 141–43, 145, 158–62, 163–65, 173–74; maps of, 144, 193n45; promoting nature and landscape, 129, 130–32, 158–59, 173; targeting women, xv–xvi, 20, 145, 152–53, 156–61, 163–65, 169, 173–74, 195n13, 196n28, 196n29
Agassiz, Louis, 50
agriculture, 2–3, 6, 28–29, 133
Ainsworth, Stephen, 35, 39–40
air-brake systems, 20, 104, 109, 114, 151
Aliquippa Grove (PA), 136
Allen, William Frederick, 21
American Angler, 141, 144, 161

American Express Company, 95
American Fish Culturists' Association (AFCS): and the Commission's salmon experiment, 44–45, 61–62, 69, 76; founding of, 43–44, 183n48; goals of, 43–45, 64; lobbying of, 44–45, 61–62; on Roosevelt's congressional speech, 65; Theodatus Garlick's contributions to, 34
amusement parks, 136–37
The Angler's Guide and Tourist's Gazetteer (Harris), 141
aquarium cars, 103–7, 120, 121. *See also* fish cars
Atchison, Topeka & Santa Fe Railroad, 126–27, 129, 136, 154–55
Atkins, Charles G., 69, 82, 183n41
Atwood, Nathaniel E., 40, 48
Audubon, James, 49
Austin, Francis, 128, 135
automobiles, 140, 174

Baird, Lydia Biddle, 49
Baird, Samuel, 49
Baird, Spencer Fullerton: and the American Fish Culturists' Association, 45, 61; and the aquarium car accident, 106; background of, 49–51;

211

Baird, Spencer Fullerton (*cont.*)
  carp experiment, 90–91, 189n38; Clackamas River hatchery, 85; death of, 188n33; as first fish commissioner, xiv, 55, 57, 181n8, 189n38; in the line fishing-trapping debate, 48, 51–52, 57–60; loaning fish cars to states, 118; lobbying of, 51–54, 63, 66; overfishing study, 57–60; pivoting to fish culture project, 60–62, 63, 66; program goals of, xiv, 53, 57, 69, 72, 185n5; rainbow trout experiment, 86, 87; recognizing railroads' support, 74, 95–97, 114–15; rejecting Green-Roosevelt proposal, 184n15; salmon experiment, 69–70, 76–77; shad experiment, 70, 71, 72, 74–76; shipping rates negotiated by, 110, 115; at the Smithsonian, 50–51
Baltimore & Ohio Railroad (B&O): assisting the Great Experiment, 96–97; building Commission fish cars, 113–14; creation of, 2–3; expansion of, 4, 7; fish stocking activities, 97, 149; promotional excursion trips, 130–31; rates charged by, 96–97, 99, 110; recognition of, 96–97, 99
Baltimore & Potomac Railroad, 99
Bangor & Aroostook Railroad, 163
bass, 93, 104, 146–47, 149, 185n24
Benjamin, John F., 53, 57, 61, 62
Bertram, James, 24
"Best Friend of Charleston" (steam locomotive), 3
Big Four, 11, 15
Boston & Worcester Railroad, 5
Bowles, B. F., 43, 44

Brackett, Mr., 82
Bradley, Daniel, 27–28
Brown, Jenn Corrine, 160
brown trout, 94, 189n45
Burlington & Quincy Railroad, 82, 110, 154
Burns, Ken, 139–40

Caldwell, Charles, 177n4
Caledonia (NY) hatchery, 34–35, 182n37, 188n28
Calhoun, John C., 177n6
California State Fish Commission, 71, 78, 93–94, 103–7, 120, 185n3
California Zephyr, 132, 192n16
Camden & Amboy Rail Road Transportation Company, 3, 21
campgrounds, 136–37
Canada, 38, 82, 183n41
canals, 1–2, 177n6
Carlson, Rachel, 174
Carnegie, Andrew, 135, 152
carp: arrival in America, 90, 189n38, 189n45; in the "Great Experiment," 90–94; impact on native species, 91, 92; public opinion of, 91, 92, 113, 189n42; railroads stocking, 149; transplanting results, 92–94, 111–12, 171, 189n35; transported in No. 1 fish car, 110–13
Cassett, A. J., 95
Catskill Rivers (Austin), 128
Cauthorn, F., 160
census (U.S.), 25
Central Pacific Railroad Company: and the aquarium car, 103–4; completion of, 12–13; construction of, 9–13; expansion of, 83; funding for, 9–10,

212 INDEX

11; incorporation of, 10–11; rates charged by, 110; ridership statistics, 13
Chandler, Zachariah, 54
Chesapeake & Ohio Railroad, 126
Chicago, Burlington & Quincy Railroad, 82, 110, 154
Chicago & Alton Railroad, 152, 153, 190n58
Chicago & Northwestern Railroad, 105, 129, 142, 143, 156, 157
Chilton, Mary Jane, 195n12
Chinese laborers, 11
Cincinnati Southern Railroad, 195n12
Clark, Frank, 115
Clark, Nelson W., 40
Clemens, Samuel, 158
Cleveland, Grover, 188n33
Clift, William, 43, 44, 61, 65, 70, 73–74
Cold Springs Trout Ponds, 37–38, 77
Collins, A. S., 35, 43, 44–45, 182n37
Colorado: Harvey depot cafes in, 154; railroad advertising, 144, 164; state fish commission, 41; as tourist destination, 127, 142
Columbia River, 76, 187n23
Commission (U.S. Fish Commission): carp experiment, 90–92, 112–13; cooperating with state commissions, 42, 61, 62, 112, 119; creation of, xiv, 47, 52–55, 62–66, 181n8; first commissioner of, 55; fish car fleet, 103, 113–16, 123; Fish Car No. 1, 108–13; fish cars (California aquarium car), 104–7; funding for, 57, 62, 64–66; goals and objectives of, 52, 66–67, 69, 119; Great Experiment of, 67, 69–70, 90–92, 171, 185n2; impact on sport fishing, 172; impacts on native species and environment, 73, 91, 171; line vs. trap fishing dispute, 47–48, 57–60; under Marshall McDonald, 188n31, 188n33; operational statistics, 116–19, 123; pivoting of, 60–62; post-1900s, 174; public opinion, 91, 106, 107, 110–13; railroads assisting, 16, 47, 76, 94–100, 172; railroads depositing fish, 95, 97–99, 108, 148–49, 171; rainbow trout experiment, 86–89, 188n31; rates charged by railroads, 94–95, 97, 99, 109–10, 112, 115; recognizing railroads' support, 95–98, 114–15; resolution creating, 55–57, 88; restocking approach, 73–74; results of, 76–77, 89, 92–94, 171; Robert Barnwell Roosevelt's role in, 62–66; salmon experiments, 76–85, 89, 185n2; shad project, 71–76, 89; transportation methods, 103, 116, 119. *See also* fish hatcheries; government (federal)
Congress (U.S.). *See* government (federal)
Conkling, Roscoe, 54
Connecticut, 24, 40, 48, 59–60, 84, 116
conservation movement: early approaches to, 26, 27, 30, 60; economic progress challenging, 28–29; impact of railroad-commission partnership on, 125; non-native species impacts, 91, 92, 185n24; railroad industry's impact on, xv, 130, 132–34, 138–39, 140, 167, 172–73; sport fishing in, 27, 125, 167–68, 193n46; women in, 167–68
Cooke, Jay, 15, 179n33
Corbett, H. W., 66

INDEX 213

Coste, Victor, 33, 34, 79
Council Bluffs (IA), 9
Credit Mobilier, 11, 13–14, 184n4
Crocker, Charles, 11
Crosby, Cornelia "Fly Rod," 162–63, 168
Cruikshank, Mrs. James A., 163
Cumberland Valley Railroad Company, 157
Current River (MO), 146–47, 148

Dawes, Henry L., 51–52, 54, 184n4
day coaches, 152
Delaware, Lackawanna & Western Railroad, 134–35, 158–59
Delaware Water Gap, 135
Delmonico (Pullman car), 20, 153
Del Monte hotel (Monterey Peninsula, CA), 126, 135–36
Denver & Rio Grande Railroad, 147, 148
Detroit, Lansing & Northern Railroad, 146
Dey, Peter A., 11
Dickinson College (Carlisle, PA), 49–50
dining cars, 19–20, 153–54, 179n29
Dodge, Grenville M., 12
Dolph, Joseph, 85
Domesticated Trout (Stone), 39, 186n14
Douglas, Stephen A., 7, 178n15
Druid Hills hatchery (Baltimore), 87
Durant, Thomas C., 12

Eastman, Frank, 113, 191n12
Edmunds, George Franklin, 53–54, 60–61, 66, 181n8
Edmunds, M. C., 43, 44, 53–54, 60–61, 69
Encyclopedia Britannica, 191n2

Erie Canal, 2, 3
Erie Railroad Company, 73, 122–23, 134
Eurasian tree sparrow, 189n46
Europe, 6, 31–33, 34, 44, 90, 127
excursion trains, 130–31, 147

Fargo, William G., 95
Farnsworth, John F., 52–53, 61
Federal Limited, 116
Ferguson, T. B., 87
financial markets, 6, 15–17, 179n33
Fish and Wildlife Service, 174, 185n24. *See also* Commission (U.S. Fish Commission)
fish cars: accidents, 105–7; building costs, 114; California aquarium car, 103–7; Commission's growing fleet of, 113–16; Fish Car No. 1, 108–13; impact of, 108–9, 113, 123; inaugural 1873 journey, 103–7; innovations in, 103, 109, 115–16; need for, 91; operational statistics, 116–19, 123; public opinion of, 112–13; retirement of, 119; for state commissions, 119–23
Fish Commission (U.S.). *See* Commission (U.S. Fish Commission)
fish commissions (state). *See* state fish commissions
fish culture: addressing declining fish stock, 21–22, 23, 26–27, 45–46, 185n27; beginnings in America, 33–40, 45–46; in Canada, 183n41; carp experiments, 90–92, 112–13; compared to fish husbandry, 30–31; and the conservation movement, 46; defined, xiii, xvii, 30–31; development of, 23, 66–67; early government interventions in, 38, 40–43;

in Europe, 31–33, 34, 44, 45; experiment results, 76–77, 89, 92–94, 171; in France, 31–33, 34, 182n28; in Germany, 31, 90; harvesting process, 79, 86; hybridization in, 23, 30–31, 35–36, 42, 183n53, 188n34; impacts on native species, 73, 91, 92, 185n24, 189n47; nationalism in development of, 44, 182n28; salmon experiments, 77–85; in Scotland, 182n27; shad experiments, 71–76, 89; shipping methods, 39, 72–73, 74, 186n14; trout projects, 86–89, 188n31, 188n34. *See also* Commission (U.S. Fish Commission); fish stocks

fish hatcheries: building and expansion of, 84, 116; Caledonia (NY), 34–35, 182n37, 188n28; Clackamas River, 85; Druid Hills, 87; McCloud (CA) salmon, 78–80, 83, 84–85, 88–89, 104, 187n22; McCloud (CA) trout, 86, 88–89; in Michigan, 40, 88, 115, 187n19; in Oregon, 187n20; Penobscot River, 69–70; role of, 22; Washington DC, 89; Wytheville (VA), 88, 188n31

Fish Hatching and Fish Catching (Roosevelt and Green), 37

fishing (excursion) trains, 147

fish stocks: commission addressing declines in, 55–60; decline due to industrialization, 26–27, 28–29, 58, 78, 83, 181n18; decline due to overharvesting, xiv, 27–28, 47–48, 51–52, 58, 171; decline of, 23–25; fish culture addressing declines in, 21–22, 26–27, 30–40, 45–46, 185n27; as food source, xiii–xiv, 23–25, 57, 149; harvesting methods, 47–48; impacts of non-native species, 73, 91, 92, 185n24, 189n47; states addressing declines in, 26, 38, 40–43. *See also* fish culture; *individual fish types*

Flint & Pere Marquette Railway, 149

fly-fishing, 160–62, 163–66, 168

"Fly Rod's Notebook" column (Cornelia Crosby), 162

Ford, Henry C., 122–23

Forest and Stream, 30, 35, 94, 121, 143, 146, 160, 168

Forest Lodge "Clubhouse" (MO), 137

forests (logging industry), 28–29, 146

France, 31–33, 34, 44, 182n28

Frisco Railroad (St. Louis-San Francisco Railroad), 97, 137, 146, 148; Forest Lodge "Clubhouse" (MO), 137

Garfield, James A., 14, 62–63, 66, 108

Garlick, Theodatus, 33–34

Gasconade River (MO), 137, 148

Géhin, Antoine, 31–32

George II (king of England), 31

Germany, 31, 44, 90–91, 94

Glacier National Park, 15, 164

Golden Spike, 12, 193n41

government (federal): Act of 1864, 10, 15; anti-railroad sentiment in, 17; creating the U.S. Fish Commission, 52–57; Credit Mobilier defrauding, 13–14, 184n4; funding the Commission, 60–62; land grants of, 7–8, 10, 15, 129, 179n28; in the line vs.trap fishing debate, 51–52; lobbying for fish culture in, 45, 52–55, 61–62; Pacific Railway Act (1862), 9–10; railroad regulations, 16–17; regulating overfishing, 59–60; Standard Time Act, 21;

INDEX 215

government (federal) (*cont.*)
states' inadequacies governing fish stock, 54–55, 59–60, 64, 71; and the transcontinental railroad, 9–10, 12, 13–14. *See also* Commission (U.S. Fish Commission)
government (state): addressing declining fish stocks, 38, 40–43; fishing legislation and regulations, 26, 29–30, 48, 59–60, 183n52; funding railroads, 5–6; inadequacies governing fish stocks, 54–55, 59–60, 64, 71; railroad pricing regulations, 16–17. *See also* state fish commissions
Grand Rapids & Indiana Railroad, 137, 142, 143–44
Grand Trunk Railway (Canada), 141
Granite Railway (Massachusetts), 2
Grant, Ulysses S., 14, 54, 193n41
grayling, 137, 142, 144, 145–46, 193n46
Great Northern Railway, 15, 164. *See also* Northern Pacific Railroad
Green, Myron, 79, 104–5
Green, Seth: background of, 34; Caledonia hatchery, 34–35, 182n37, 188n28; cross-breeding experiments of, 35–36, 42; impact on fish culture, 34–37; and Robert Barnwell Roosevelt, 36–37, 61, 184n15; shad projects, 36, 41, 70, 71, 72, 74; and Stephen Ainsworth, 35, 40; trout projects, 87, 186n14
guidebooks, 132, 141–43, 146
Guidebook to the Fishing and Hunting Resorts (guidebook), 141

Hallock, Charles, 29
Halverson, Anders, 89
Harlan, James, 18
Harper's Bazaar, 160, 195n16
Harper's New Monthly Magazine, 47, 127
Harper's Weekly, 131, 153, 155
Harris, William C., 141, 185n27
Harvey, Fred, 154–55
Harvey & Company, 158
Harvey Girls, 154–55
Hassam, Frederick Childe, 131
Hessel, Rudolph, 90–91, 189n38
Holberton, Wakeman, 87
Hood, J. H., 99
Hopkins, Mark, 11
hotel cars, 19–20, 153–54
Hotel Del Monte (Monterey Peninsula, CA), 126, 135–36
Howe, Timothy, 54
Huningue (Alsace, France) hatchery, 32–33
Huntington, Collis P., 11, 12
Huntington, J. D., 43
husbandry. *See* fish culture

Idlewild Park (PA), 136
Illinois Central Railroad, 7–8, 110, 122
Illinois State Fish Commission, 117–18, 122
industrialization, 28, 29, 47, 58, 78, 181n18
Inness, George, 130
inspection locomotives, 149
Instructions Practiques sur la Pisciculture (Coste), 33
Interstate Commerce Act (ICA), 17
Interstate Commerce Commission (ICC), 17
In the Maine Woods, 163
Iowa State Fish Commission, 41, 120, 122

Jackson, Lansing & Saginaw Railroad, 145
Jacobi, Stephen Ludwig, 31
Japan, 93, 94
Jay Cooke & Company, 15, 179n33
Jerome, George H., 28, 98, 190n54
Jim Crow passenger cars, 195n12
J. P. Morgan, 17, 179n33
Judah, Theodore, 10–11

Kansas Pacific Railroad, 74, 154–55
Kansas State Fish Commission, 41, 191n22
Keeler, Sanford, 149
Keyser, William, 99
Kimble, Arthur Reed, 132–33
King, John, 99, 178n15
King, William R., 7

Lackawanna Railroad, 134–35, 158–59
The Lackawanna Valley (painting), 130
ladies cars, 157–58
Ladies Home Journal, 159
land grants, 7–8, 10, 15, 129, 179n28
land seeker tickets, 129
Ligonier Valley Rail Road, 136
Lincoln, Abraham, 18, 178n15
line fishing, 47–48, 51–52, 58
Literary Digest, 164, 196n27
locomotives, 3, 18–19, 149
logging industry, 28–29, 146
London & North Eastern Railway, 145
Ludlow, William H., 66–67

Magazine of Hanover, 31
Maine: Acadia National Park, 140; fishing regulations in, 183n52; railroad advertising promoting, 162–63; salmon hatchery, 69–70; state fish commission, 40

Maine Central Railroad, 140, 141, 143, 147, 162–63
Man and Nature (Marsh), 26
Manual of Railroads, 177n2
Marbury, Mary Orvis, 161, 168
Marsh, George Perkins, 25–27, 28, 29, 47, 50, 181n8
Maryland: canals in, 2; fish stocking in, 84, 87, 99, 149; hatcheries in, 109, 115; state fish commission, 99
Massachusetts: addressing declining fish stocks, 40, 41, 48, 51–52; early railroads in, 2, 4; regulations, 48, 59–60; state fish commission, 38, 40, 41
Mather, Fred, 40, 43, 183n48
Mauch Chunk railroad, 2
McBride, Sara Jane, 168
McCloud River salmon hatchery, 78–80, 83, 84–85, 104, 187n22
McCloud River trout hatchery, 86, 88–89
McDonald, Marshall, 85, 88–89, 111, 112, 113, 188n31, 188n33
messengers (fish): challenges faced by, 73; and fish cars, 103–4, 108, 109, 114; operational statistics, 116–19, 123; railroad fees, 95, 96–97, 99, 101, 110, 115; transporting fish cans, 72–75, 83–84, 95
Michigan: conservation groups in, 193n46; early railroads in, 3; federal fish hatcheries in, 40, 88, 115; fish stocking in, 84; increased tourism and sport fishing, 137, 143–44, 145–46; overfishing in, 28, 29, 54
Michigan, Lake, 53, 54, 58, 84, 178n20, 187n19

INDEX   217

Michigan State Fish Commission: fish cars used by, 120, 121; founding of, 41; objectives of, 42, 54; on overfishing, 28; recognizing railroad contributions, 98; restocking programs, 98; salmon state hatchery, 40, 187n19
Miller platform, 104, 109, 114
Miramichi River salmon hatchery, 38
Mississippi River, 61, 63, 70, 71–72, 78–79
Missouri Pacific Railway, 144
Missouri State Fish Commission, 92, 99–100, 121, 190n58
Mitchell, John, 85
Mobile & Ohio Railroad, 7, 144
Mohawk & Hudson Railroad, 3
Monon Route, 193n45
Montague, Mr., 78
Montana State Fish Commission, 191n22
Mount Washington Railway, 128
Muir, John, 133, 139
Munsey's Magazine, 164, 196n28

The National Parks (documentary series), 139–40
national park system, 138–40, 167, 192n31, 193n33
Native Americans, 12, 78, 179n28, 186n12, 187n22
Nebraska State Fish Commission, 88, 119–20, 121
New Harper's Monthly Magazine, 131
New Jersey, 4, 40, 84
New York: early railroads in, 4; fishing regulations, 59–60; recognizing railroads' support, 99; underwriting railroads, 178n11

New York & Erie Railroad, 6
New York Central & Hudson River Railroad, 99, 157
New York Central Railroad, 105, 132–33
New York Sleeper Car Company, 152
New York Sporting Goods Company, 166
New York State Fish Commission: establishment of, 40; and the federal fish commission, 45, 62, 112; fish cars used by, 121; fish stocking programs, 71, 112, 188n28; hatcheries of, 182n37, 188n28; recognizing railroads, 99
New York Times, xiv, 13, 112, 138, 157, 189n42
Norris, Philetus W., 91
North and West Illustrated (railroad guidebook), 142
Northern Central Railroad, 99
Northern Pacific Railroad: advertising of, 142, 164; bankruptcy of, 179n33; and the creation of Yellowstone National Park, 138; dining cars, 154; fish stocking activities of, 148; land grants, 15; specialized hunting-fishing cars, 147–48; westward expansion, 15
Northwestern Pacific Railroad, 145, 193n45

observation cars, 131
Official Railway Guide of the United States and Canada, 21
Ohio State Fish Commission, 41, 121
Omaha (NE), 9, 11–12
Opelousas Courier, 112
Oquossoc Angling Association, 183n52
Oregon, 85, 187n20

Ornithological and Piscatorial Acclimatization Society of California, 188n28
Orsi, Richard, 133
Orvis, Charles F., 161
Orvis Company, 161, 166
Osgood, Edward, 104–5
Outing Magazine, 161
The Outlook, 132–33
Overland Stage Express Company, 179n30

Pacific Grove campground, 136
Pacific Railway Act (1862), 9–10, 18
Page, George Shephard, 45, 62, 63, 183n52
palace cars (Pullman), 152–53
passenger cars: day coaches, 152; dining cars, 19–20, 153–54, 179n29; hotel cars, 19–20, 153–54; Jim Crow cars, 195n12; ladies cars, 157–58; observation cars, 131; palace cars, 152–53; sleeper cars, 19–20, 152–53, 179n29; technological innovations in, 19–20; tourist cars, 153
Pearline laundry soap, 159
Pennsylvania: canals in, 2; early railroads in, 4; fish stocking in, 84; state fish commission, 40, 121, 122–23; tourism in, 134–35, 136; underwriting railroads, 178n11
Pennsylvania Central Railroad, 6–7, 96
Pennsylvania Railroad Company, 95, 96, 109, 110, 141, 143
Penobscot River salmon hatchery (Maine), 69–70
Pere Marquette Railroad, 149
Perrin, William T., 79, 104–5

Perrysburg Journal, 106
Philadelphia, Wilmington & Baltimore Railroad, 96, 109–10, 157
Philadelphia & Reading Railway Company, 157–58
Pittsburgh & Lake Erie Railroad, 136–37
Pocono Mountains, 134–35
Poor, Henry V., 177n2
Poppe, Julius A., 189n35
Promontory Summit (Utah Territory), 12
propagation. *See* fish culture
Pulitzer, Joseph, 135
Pullman, George, 152
Pullman Company: advertising targeting women, 156, 164–65, 196n29; dining cars, 19–20; hotel cars of, 19, 153–54; improvements by, 19–20; palace cars, 152–53, 157; specialized hunting and fishing cars, 148

rail gauge, 17–18, 114, 180n40, 180n41
railroad commissions (state), 16
railroad industry: assisting the Commission, 43, 65, 76, 83–84, 94–101, 111–12, 116, 123, 125, 171–72; Commission recognizing, 74, 96–98, 108, 114–15, 118, 190n58; early years (1830 to 1860), 1–8; economic impacts of, 5–7, 8, 17, 179n33; employees distributing fish, 73, 80, 95, 97–99, 123, 148–49; fuel used, 158; funding, 5–6, 178n11, 180n37; innovations in service, 151–55; innovations in technology, 17–21; post-1900s, 174; in the post-Civil War era, 9–17; public opinion of, 15–17, 111–12, 133, 184n4, 192n20, 193n32; racist policies, 195n12; "railroad gardening"

INDEX 219

railroad industry (*cont.*)
movement, 133; rates for passenger travel, 128–29, 179n29; rates for shipping, 16, 94–95, 96–97, 109–10, 112, 116–17, 118, 122–23; rates for specialized fishing/hunting cars, 148; revenue sources, 101, 126–28, 145–46, 179n29; role in conservation, xv, 125, 130, 132–34, 138–39, 140, 167, 172–73; role in national parks creation, 138–40; role in sport fishing, 101, 125, 145–48, 150, 160–61, 172–73; scandals of, 13–14, 15–17, 179n33, 184n4; and Smithsonian expeditions, 51; societal impacts of, 4–5, 8, 20–21, 156; state commissions' use of, 99–100, 119, 122–23, 125; timetables, 20–21, 128–29; and tourism, 19–20, 101, 125–28, 134–40, 145–47, 150. *See also* advertising (railroad industry); passenger cars; transcontinental railroad

Railroad Land Grant Act (1851), 7

"railroad time," 21

rainbow trout: at the Caledonia hatchery, 188n28; in the Great Experiment, 86–89, 92–94, 188n31; hybridization of, 188n34; sport fishing, 137; transplanting successes, 89, 92–94, 171

recreational fishing. *See* sport fishing

Redmond, Roland, 87

regulations: addressing overfishing, 26, 29–30, 48, 59–60, 171, 183n52; addressing railroad pricing, 16–17; AFCA lobbying for, 44–45. *See also* government (federal)

Reiger, John, 167, 181n8

Reinhart, C. S., 160

Remy, Joseph, 31–32, 182n28

Rhode Island, 48, 59–60

Rhode Island State Fish Commission, 40

rights-of-way (railroad), 133, 167

Riley, Glenda, 167, 168

Robertson, E. W., 112

Rock Island System, 145, 164

Rod and Gun, 160

Roosevelt, Robert Barnwell: advocating for fish culture, 23, 36–37, 61, 185n24; on declining fish numbers, 27; role in creating U.S. Fish Commission, 62–66; and Seth Green, 36–37, 61, 184n15

Runte, Alfred, 133, 140

Rutland & Burlington Railroad, 128

Sacramento River, 77–79, 83

safety (railroad travel), 19, 20, 104, 151, 156, 157–58, 169, 173

Salem & Little Rock Railroad, 146–47, 148

salmon: Commission's East Coast experiments, 69–70, 185n2; declining numbers of, 24, 28, 78, 187n23; early attempts to propagate, 37–39, 42, 183n41; in Germany, 31; in the Great Experiment, 70, 76–85; harvesting process, 79; hybridization of, 36; impact on native species, 185n24; in the McCloud River basin, 187n22; methods of harvesting, 47–48; methods of packing, 79–80; Michigan fish commission transplants, 187n19; proposed restocking projects, 45, 60; results of experiments, 80–81, 83–85, 92–94, 171; role of AFCA in propagation of, 44; shipping challenges, 79–83

Sandiford, Glenn, 189n38

Sanford, Henry, 96
Santa Fe (NM), 126–27, 136
schedules (railroad timetables), 20–21, 128–29
Schlict's Mill (MO), 137
Schullery, Paul, 30, 132, 138, 161
Scotland, 41, 94, 182n27, 189n45
shad: in Commission projects, 69, 70, 71–76; declining population, 25, 28; as dietary staple, 71; early attempts to propagate, 42, 45; impact on native species, 73, 185n24; methods of harvesting, 47–48; natural habitats of, 71; proposed restocking projects, 61, 63, 66; role of AFCA in propagation of, 44; Seth Green's early experiments with, 36; shipping challenges, 72–73, 74
Shaffer, Marguerite S., 139
Shanks, W. F. G., 23–24, 181n18
Shaw, Mr. (Drumlaning, Scotland), 182n27
Shoemaker, Samuel M., 96
Sierra Club, 139
6000 Miles Through Wonderland (pamphlet), 142
Slack, J. F., 43
sleeper cars, 19–20, 152–53, 179n29
Smithsonian Institute, 50–51, 57
Snow, Phoebe, 158–59, 195n13
South Carolina Canal & Rail Road Company, 3
Southern Pacific Railroad: amenities offered by, 136; assisting the Commission, 78; formation of, 15; hotels of, 126; impact on tourism, 126, 129, 135–36; role in conservation movement, 133, 139

South Side Club (New York City), 87
Spokane & Inland Empire Railroad, 143
sport fishing: catch-and-release fishing, 141; catering to women, 166; class privilege in, 27, 30, 165–66, 194n8; and the conservation movement, 27, 125, 167–68, 193n46; early fishing clubs, 30; "fishing line" advertising campaigns, 143–45; fly-fishing, 160–62, 163–66, 168; overfishing, 27–28; railroad advertising influencing, xv–xvi, 20, 141–43, 145, 158–62, 163–65, 173–74; railroad-Commission projects impacting, 123, 125, 172–73; railroads democratizing, 134–35, 150; railroads' impact on, xv–xvi, 101, 128, 137, 145–47, 151; as socially acceptable for women, xv–xvi, 135, 145, 151, 159–60, 168–69, 173–74; specialized train cars for, 147–48; women as experts in, 161–63
Sportsman's Exposition (1895, Madison Square Garden), 162
Standard Time Act (1918), 21
Stanford, Leland, 11, 12
state fish commissions: addressing declining fish stocks, 38, 40–43; carp programs, 91–92; cooperating with federal commission, 42, 61, 62, 104–7, 112, 119, 125; establishment of, 40–43; fish cars used by, 119–23; obstacles face by, 41–42; railroad shipping fees, 16, 122–23; railroads providing support, 74, 99–100, 119, 122–23, 125; rainbow trout programs, 87–88, 89; recognizing railroad support, 99, 190n58; use of federal fish cars, 117–18. *See also* government (state); *individual states*

Stilgoe, John, 128
St. Louis, Iron Mountain & Southern Railroad, 195n12
St. Louis-San Francisco Railroad, 97, 137, 146, 148
Stone, Livingston: in the American Fish Culturists Association, 43, 44; background of, 37; California aquarium car trips, 104–7; Cold Springs Trout Ponds hatchery, 37–38; as deputy fish commissioner, 72, 77; end of career with Commission, 188n33; experiment results, 72, 84–85; impact on fish culture, 37–39; rainbow trout experiment, 86–88, 89; salmon experiment, 72, 77–85; on Seth Green's success, 35, 36; and the Wintu, 186n12, 187n22
Stover, John, 16
The Sun, 14
Sundry Civil Appropriations Act (1872), 66, 69
Susquehanna River, 80–81

telegraph, 178n17
Texas and Pacific Railroad, 15
Texas State Fish Commission, 41, 111–13
Throckmorton, Mr., 106
Thurman, Allen, 54
Time-Table Convention and the General Time Convention, 21
timetables (railroad schedules), 20–21, 128–29
time zones, 21
Tomlin, Lucy, 160
tourism: beginnings of, 9, 125–26; defined, 191n2; impact of changing social norms on, 4–5, 151, 159, 169; railroad advertising encouraging, xv–xvi, 127, 128–32, 135–36, 141–43; railroad innovations impacting, 19–20, 151–55; railroad-owned lodging, 135–37; railroad revenue from, 101, 127–28, 179n29; railroads repurposing old lines for, 145–47; role of railroads in growth of, 125–28, 134–40, 150. *See also* sport fishing
tourist cars (Pullman), 153
Townsend, Mary Trowbridge, 161, 166
track gauge, 17–18, 114, 180n40, 180n41
transcontinental railroad: benefits of, 13, 126, 186n11; completion of, 12–13; construction of, 9–13; early proposals for, 9, 178n20; first bicoastal fish transplant on, 71; funding for, 9–10, 11, 129; impact on tourism, 126, 129, 135–36, 140; legislation enabling, 9–10; mail delivery on, 13, 179n30; national interest in, 12–13; passenger rates, 110, 179n29; pro-Northern bias in, 15, 180n40; role in fish culture, 65, 71; scandals impacting, 13–14; southern routes, 15. *See also* railroad industry
transportation: automobiles, 140, 174; pre-railroad options, 1–2; trucking, 119, 174; water transportation routes, 1–2, 177n6. *See also* fish cars; railroad industry
trap fishing, 47–48, 51–52, 58–59
A Treatise on the Propagation of Certain Kinds of Fish (Garlick), 34
trout: at the Caledonia hatchery, 188n28; California commissioners interest in, 185n3; Commission's approach to stocking, 188n31; declining numbers of, 27, 29; in

early private preserves, 30; early propagation of, 33, 35–36; in France, 31–33; in Germany, 31; goals for propagation of, 64, 185n27; in Great Experiment projects, 86–89, 92–94; hybridization of, 36, 188n34; impact on native species, 185n24; methods of transporting, 186n14; mindset underlying stocking of, 66–67; in railroad advertising, 137, 142, 161, 164; railroads stocking, 148–49; role of the AFCA in propagation of, 44; Scottish origins, 94, 189n45; transplanting results, 92–94, 171; women fishing, 161, 164
Trout Culture (Green), 36
Trout Unlimited, 193n46
trucking, 119, 174
Tucker, Payson, 162–63
Turner, Spencer, xiii
Tuscumbia, Courtland & Decatur Railroad, 3
Twain, Mark, 158

Union Pacific Outings (railroad brochure), 142
Union Pacific Railroad Company: advertising encouraging tourism, 129; aquarium car accident, 105–7; completion of, 12–13; construction of, 9–13; Credit Mobilier scandal, 13–14, 184n4; dining cars, 154; funding for, 10, 11; land grants given to, 10, 179n28; legislation creating, 9–10; passenger statistics, 13; rates charged by, 110, 118
urbanization: declining fish stock due to, 28, 47, 58; population growth in, 25; railroads offering escape from, 4–5, 125–26, 132

Vermont, 25–27, 41–42, 181n8
Vermont State Fish Commission, 40, 41–42
Vernon, Edward, 21
Virginia, 2, 88, 116, 188n31
Virginia State Fish Commission, 41, 188n31

wagon routes, 177n6
Washington, Booker T., 108
Waterloo Courier (IA), 122
water transportation routes, 1–2, 177n6
Webster, Daniel, 5
Welsh, William, 164–65, 196n29
Western Maryland Railroad, 99
Westinghouse, George, 20
Westinghouse air brake system, 20, 104, 109, 114, 151
whitefish, 53, 58, 66, 93, 98, 190n54
White Sulphur Springs (WV), 126, 192n24
Whitney, Asa, 178n20
Wilkins, George, 99
Williams, Anna, 160
Wilmington & Baltimore Railroad, 96, 109–10, 1057
Wintu tribe, 78, 186n12, 187n22
Wisconsin State Fish Commission, 41, 121, 122
women: advertising highlighting safety, 20, 151, 156, 157–58, 173–74; advertising of sport fishing for, xv–xvi, 145, 151, 159–63, 164–66, 173–74; advertising showing travel comfort, 152–53, 173–74; changing social norms for,

women (*cont.*)

    xv, 20, 135, 155–56, 159–60, 161, 169, 195n16; and the conservation movement, 167–68; fly-fishing by, 160–62, 163–66, 168; Phoebe Snow advertising campaigns, 158–59, 195n13; role of class privilege, 165–66, 194n8; sport fishing as socially acceptable, 135, 173–74, 196n29; sport fishing contributions of, 162–63, 168–69; sport fishing industry catering to, 166

women's clubs, 168
Woodruff, Theodore T., 152
Woods Hole (MA), 51, 57, 116
Wyoming State Fish Commission, 41, 42
Wytheville (VA) hatchery, 88, 188n31

Yellowstone National Park, 91, 138, 140, 142, 161, 167–68
Yosemite National Park, 139, 168

www.ingramcontent.com/pod-product-compliance
Lightning Source LLC
Chambersburg PA
CBHW021853230426
43671CB00006B/373